Opening Up

Volume I: The Making of a Surgeon

David Clarke

McGeary Media

Preface

This volume, together with its successor, is an edited and shortened version of a memoir I wrote shortly after my retirement as a consultant surgeon. In longhand and then typed laboriously with one finger, it came to 220,000 words and was originally created for family.

The motives for a memoir are various, but all driven by the fact that after one departs this life, the knowledge and experience of a lifetime will be lost forever, unless it is recorded. In my case, I have led what to me has been an interesting life, enriched by an association with a wide variety of characters and I have been fortunate in having a good memory, evidenced by the comments of several of my contemporaries over the years ('You should write a book, David.')

In collaboration with the journalist Michael McGeary (The Memoir Man), the original manuscript was published privately. Naturally, it contained much detail of a personal nature that would have little or no interest to the general reader. However, Michael persuaded me that a shortened version might do so and this volume is the result. I am extremely grateful to him as a colleague and friend for his skilled assistance in this process.

During my time there have been more advances in medicine than since the time of Hippocrates, and I hope the historical digressions will prove informative and/or entertaining.

Finally, despite their associations with disease and death (most of us no longer die at home), in my day hospitals were happy places in which to work, with great camaraderie. And I hope this has been successfully conveyed in this narrative.

David Clarke, June 2023

Contents

Prologue

August 1st, 1966. At eight o'clock in the morning, after five years of study, twenty-four newly qualified doctors assembled in the magnificent board room in the Royal Victoria Infirmary (RVI) in Newcastle-upon-Tyne. Seated around the long, mahogany table and surrounded by oil paintings of long-dead dignitaries, excited but somewhat apprehensive, they received a brief welcome, which oddly enough comes from one of their own, who must have been instructed earlier. They then collected their "bleeps", keys to their room and a slim handbook, and were dispersed to their respective posts. They were now termed "housemen" (including the two females), so named because their new home for the next year was the doctors' residence, or "house", next to the nurses' home, in one wing of the hospital. I was one of their number, now embarking on the first day of my career in the noble art and science of medicine. This is my story...

Chapter One

Casualties of War

One warm sunny day in May 1943, my mother wheeled my pram to East Park and sat on a bench while I slept. She was suddenly roused from a daydream by the vision of a cloud. Coming out of the cloud and towards her like a train emerging from a tunnel was my father's face.

'It was a bomb, Olive,' he said.

Shaken, she set off home immediately, and as soon as she entered the house, there was the dreaded telegram on the tablecloth.

A member of a gun crew, Dad was struck by fragments of an Italian bomb that landed close to his ship on May 14th 1943. Two other members of the crew were killed, and they were all buried at sea after the ship sailed the same day. I was six months old, and he was twenty-three. He had never set eyes on me.

I was born at five-thirty in the afternoon of Thursday November 26th 1942, at 3 Ivy Terrace, Barnsley Street, in the proudly named City and County of Kingston-upon-Hull. My parents married on Easter Sunday 1941 at Holy Trinity Church, the largest parish church in England, in the historic Old Town. At the time of my birth, my father was at sea as an Able Seaman on the destroyer *HMS Ilex*. The house we shared with my mother's parents had already sustained damage three times—twice just to the roof, and once when an incendiary fell straight through the living room ceiling and crushed my father's trilby, which was on a chair. Apparently, the fire was easily extinguished. My father, who was on leave, joked he would rather be at sea, as it was safer. Sadly, this turned out not to be the case.

The trilby incident may have occurred about the time that Uncle Tom was killed, along with fifty-two others, by German bombers on the night of July 18th, 1941. Uncle Tom's last words were, 'Don't look back, May, your house has gone!' Seconds later, so had he.

Years afterwards, I learned that when the family went to what remained of Mulgrave Street the following morning to see what they could salvage, the house had been looted. I was shocked to the core. Even in middle age, I was still naive enough to believe in the so-called Dunkirk spirit and found it difficult to believe that neighbours would do such a thing. We live and learn.

Aunt Elsie (known to her husband by her middle name, May) and Tom's only child, Stanley, was about seventeen. Tom was forty-eight and had become very left-wing politically because of his experiences during service in World War One. It was said that he came downstairs each morning with a salute and the greeting, 'Morning comrades!' He advised his son to join the RAF when he was old enough as 'only the officers get killed', but this advice was either not heeded or Stan had no choice in the matter because he joined the Navy and served on the aircraft carrier, *HMS Victorious*.

It is said that a child can't remember anything before the age of three, but I can. I recall being lifted towards the light in the living room as a toddler and struck by both the brightness of the light and the closeness of the point where the picture rails joined at the upper corner of the room as I was bounced around. Until I was about four, the light was an incandescent white light provided by coal gas feeding three or four gas mantles suspended from a metal tube incorporating a chain and pulley, which somehow switched on the supply. As a toddler, the fragile white tracery of these mantles fascinated me as I saw them close up when one needed changing; they were like the skeleton of an exotic marine creature. I also distinctly remember toddling in the dark from our house to the Anderson shelter opposite, looking up at the searchlights, hearing the German bombers, entering the dark shelter with my mother and grandparents and, vaguely, of seeing the man who lived in the house beside it. For many years afterwards I could recall the distinctive smell of this air-raid shelter, an amalgam of musty mould, damp concrete and the stale beer refreshing the man opposite. Whenever in later life I encountered a similar smell (usually bombed buildings, ruins or a damp cellar), the vision of the shelter would sweep back suddenly and be quite vivid, taking me back in time. Both the first and last daylight raids on Britain were on Hull, and considering that the last was on March 17th 1945, when I was two years and four months old, it's surprising that I should have such recollection.

My mother and I lived with her parents until I was eight. The six houses in Ivy Terrace were so-called 'sham fours', two up and two down, with a tiny front garden, backyard and outside toilet. The backyard had a homemade, ramshackle construction of tarpaulin housing chickens and pullets and later rabbits. A similar lean-to construction against the rear wall of the house served as a laundry, with an old-fashioned mangle, washtub and poss stick, which was used to pound dirt out of the clothes.

Mother was Olive Evelyn, the youngest of three sisters; the others, Elsie May Busby and Emma Florence Gunson, being eighteen and thirteen years older, respectively. Before her marriage, my mother worked as a sales assistant at Edwin Davis, a department store in the city centre (bombed out during the war), and it was there that she met my father, George Henry Clarke, three years younger than her. However, before being conscripted, his occupation is listed as a cleaner at Joseph Rank's Clarence flour mill near Drypool Bridge over the river, where his father, Charles Clarence, was a miller. From besuited salesman to dusty cleaner seems a step backwards, but it was almost certainly because Edwin Davis was bombed, so he lost his job there.

On November 18th 1940 he left for the training ship, the shore-based *HMS Ganges*, as an ordinary seaman, then to *Victory 1* on January 31st 1941 and, finally, as an able seaman to his first ship, *HMS Veronica* (also known by its pennant number, K37), a corvette of the Flower class. She was built at Smith's Dock in Middlesbrough for the French Navy, but the Royal Navy took her over after the fall of France. Launched in October 1940, she acted as an escort for Atlantic Convoys, which involved calling at Iceland, and she also took part in the escort of *HMS Prince of Wales*, which took Winston Churchill to his first meeting with President Roosevelt at Placentia Bay, Newfoundland, from August 8th to 11th 1941. My father was on *Veronica* at that time, but I doubt if he would have known the significance, other than that a battleship would not normally form part of a convoy. He was then transferred to the destroyer *HMS Ilex* (D61), launched in 1937. I have a small calendar for 1942, on the back of which he has noted various ports of call after sailing on July 18th 1942—New York; Charleston, South Carolina; Key West, Florida; Guantanamo; San Juan and, finally at anchor, Mayaguez, Puerto Rico. I do not know what the purpose of this long voyage was, but the ship then appears to have sailed across the Atlantic to Freetown in Sierra Leone. The next set of postcards shows Gibraltar, Ain El Turk, Algiers, and most poignantly, Oran, in Algeria, where he was killed. Two of his shipmates came to visit my mother with a collection of cash from the crew when they were next on leave and she never forgot their kindness because neither lived locally.

My mother's father, Granddad Barmby, was the biggest influence on my early life, as he did his best to be a surrogate father. He could read and write, but otherwise was poorly educated, and he had always been in unskilled, poorly paid work, initially as a farm labourer near his birthplace, Fraisthorpe, in the East Riding, and later, after migrating to Hull, as a council worker, with a horse and cart (or *rully*, to use the local term). To me, he will always be a gentleman. He gave me that precious commodity, time, and was unfailingly patient, kind and loving. He was a Labour man, reading the *Daily Herald* and being a member of the Transport and General Workers' Union. I have photos of him at a TGWU convalescent home after a bout of pneumonia, which he suffered at least twice. Hull East has always been one of Labour's safest seats. Commander Harry Pursey was the local MP, holding his seat for twenty-five years, from 1945 to 1970, when John Prescott succeeded him.

'Vote, vote, vote for Mr Pursey, you can't vote for a better man—yes you can!' was the chant at election time. When asked about Mr Churchill, Granddad would simply say, 'He lost a lot of our lads at Gallipoli.' But he didn't really have much interest in politics. His only vices were his pipe, the chewing of old-fashioned hard, black liquorice, which you no longer see, and at Christmas, a box or two of crystalline ginger. He was teetotal, but to my consternation, I learned in later life that as a young man he had been a drinker, as so many poor men were. Apparently, one night Grandma sent Auntie Elsie, then a young girl, to fetch him from the pub. He decided it was time to stop drinking when your own daughter was given such a task, and so he did. Granddad became a Methodist and a member of the Sons of Temperance, although when I knew him he no longer worshipped. He was thirty-eight at the outbreak of World War One. Although he joined up, he remained in the UK and worked with horses in some capacity, a huge number being required on the Western Front. Granddad had an allotment about a ten-minute walk from home, alongside the Sutton drain, where we could almost guarantee hearing the plop of a water rat (as we called them then, but really voles, cute and harmless) as it entered the water and swam with its little head just visible. But Sutton drain has been filled in for years now, so the habitat has gone, and, indeed, Ratty, of *Wind in the Willows*, is now a scarce creature nationally. We frequently went to the allotment, which was partly a refuge for Granddad, and he gave me my own little patch.

Another influence on my childhood was the aftermath of the war. Hull was the most heavily bombed British city apart from London, being the third most important port, after the capital and Liverpool. Air raids began on June 19th 1940 and continued until March 1945, but were at their most intense between May 1941 and July 1943. Contemporary radio and newspaper reports did not identify the city by name, referring to it as 'a north-east town', to avoid giving tactical information to the enemy. This omission has always riled the citizens of Hull, who felt their suffering remained unknown to their compatriots compared to that of other provincial cities such as Liverpool, Coventry and Plymouth. Even now, whenever the subject is raised on television or in the press, they rarely mention Hull. In fact, the raids damaged 87,000 buildings and of the population of about 300,000 at the outbreak of war, some 152,000 were made homeless.

Air raids killed around 1,200 civilians and left 3,000 injured—relatively small numbers considering the devastating damage incurred. By the end of hostilities, only 5,945 out of 92,660 homes in Hull had escaped some form of bomb damage.

For the first few years of my life there was spare ground in Barnsley Street opposite Ivy Terrace. I remember vividly what must have been VE Day because on this site were two large school blackboards, the mobile type on casters and with wooden frames containing the board, which could swivel so they could write on both sides. One face had a chalk portrait of Hitler and the other of Mussolini, and a bonfire had been lit at their base. As they burned, a party was being held, accompanied by cheering.

Before too long, brand-new council houses with inside toilets occupied that area. Just round the corner from Barnsley Street was Dansom Lane with a haulage-cum-scrapyard company, Ulyott's, where for a while Granddad was the night watchman. He sometimes took me during the day and I played in the cockpit of the wingless fuselage of a Junkers 88, which was intact in the yard, with the large, black cross edged with white on the side of the grey tube. The next-but-one terrace from ours was totally destroyed by bombing and it took a long time for the area to be cleared. I recall playing among the ruins one hot day, using an oblique collapsed concrete slab as a roof for a den, while an older girl sunbathed on the top of it. No health and safety in those days!

Granddad was keen on rugby league and took me to my first game when I was four. We went to see Hull Kingston Rovers, the team of the east of the city, Hull RFC being that of the west. We always stood in the open at the top of the terrace at the far end. I learned to recognise all the colours of the opposing sides and some of the well-known players, like Lewis Jones of Leeds, who was bald, bandy-legged and did not look like an athlete. Some of these teams are now defunct in the modern era of full professionalism, from places such as Bramley, Hunslet and Dewsbury. Even Liverpool had a lowly team who, like Doncaster, were always near the bottom of the league. Featherstone's ground could contain the population of the town.

One day, Granddad took me to a small flat in the oddly named district of Hull called Wincomlee. It belonged to an old man called Anthony Starks, who once played for Hull KR and Great Britain. Starks sat in his armchair wearing a velvet, Victorian-style smoking cap of Turkish design with a tassel. He was smoking his pipe, with a polished brass spittoon next to him containing sand or wood shavings. Against the wall was a display cabinet with several of his international caps, all topped with gold tassels. But the most striking feature, which fascinated me, was his outstretched artificial leg. It never occurred to me that he had not always had it, and I struggled to imagine how he managed to run, let alone become a rugby international. I can only assume that Granddad knew him because they both worked for the council—in those days all the players had another job. In the *Hull Daily Mail's* centenary history of Hull KR, published in 1993, there are several photos of Anthony Starks, taken in the 1890s, a tall, strong-looking man, with a full complement of legs.

Granddad read me stories when I was in bed with measles. I loved Woodward's Gripe Water and could have become addicted to it, but hated Fenning's Fever Cure (not surprisingly, as I later discovered it was simply very dilute nitric acid, no wonder it tasted foul—it's amazing they allowed such practices) and had my nose painfully pinched every morning before a tablespoonful of Scott's Emulsion was poured down my throat. It was a thick cream-coloured liquid with a picture on the bottle of a fisherman in a sou'wester carrying a cod over his shoulder.

Shortly after I had the mumps, poor Granddad came downstairs one morning looking like a hamster with food in his cheek pouches. I had passed it on to him. It's surprising that an old man who was one of twelve children had not caught it before.

Friday night was bath night for me, as it was for the rest of the country, just as Monday was washday. We had a zinc bath, kept hung up on a nail in the scullery. I can only think in retrospect that Mother and Grandma chose to bathe when I was out, but Granddad went to the Turkish baths weekly. Each morning you could hear him at the scullery sink, washing and splashing, shaving and always washing his hair. Unless he was going out, he wore his shirt with the collar detached, and he kept his trousers up with both belt and braces. My grandparents' generation avoided waste. Small fragments of soap were collected in a cube-shaped cage with a handle and agitated in water to create a lather, so that every piece was used.

We always seemed to have roast beef and Yorkshire pudding on Sundays, the Yorkshires served first on their own, almost the size of the plate and lathered in onion gravy. The leftovers were eaten on Monday, partly for economy and partly because it was washday, so they served the meat cold with mashed potato.

I hated Mondays, especially in winter, because the washing was dried in front of the fire on clothes horses, and having just come in from the cold, I couldn't get close enough to warm up. Grandma made the most amazing piccalilli, to eat with cold meats and produced in jars holding a couple of litres or so. She also made delicious ginger beer and in the terrible winter of 1947 (how did we ever survive it, with a single coal fire?), aged five, I cleared snow from our path, tiny though I was. I was kept going by frequent doses of this potion, which suffused the extremities with heat before we had ever heard of Heineken.

I always enjoyed trips to a shop on Holderness Road that sold only tripe, cows' heel, pigs' trotters and brawn, an opalescent, brownish jelly embedded with pieces of meat, usually the cheaper cuts from a pig. This was all proletarian fare, of course, but the shop was a long-established business, so it must have made a profit. It was lined from floor to ceiling with white ceramic tiles, while around the periphery and in the windows were glass cases containing stuffed game birds, a fox, squirrel and badger. I loved looking at them while we waited.

Another cheap dish was sheep's head broth, which I adored. It was cooked with a variety of vegetables for what seemed an eternity in a large black pan on the coal fire. I took special delight in sucking the soft brain tissue through my teeth. One day Grandma asked me to, 'Go ont' road' and buy a sheep's head. When I got to the front of the queue and made my request, the butcher gave a little sigh, went to the back of the shop and returned with one, still with fleece attached and cud in its mouth (it must have died happy). He washed it clean, skinned it and removed the eyes, before finally bisecting it with a single blow of the cleaver. During this time the line behind me grew, but as I recall they did not seem too restive—people were more patient than they are now. At the end of the operation, he wrapped the two halves in newspaper and said, 'That'll be a tanner, son.' Six old pennies—two and a half pence now! Years later, with my medical knowledge, I wondered why we didn't get scrapie, the equivalent disease in sheep to CJD in cattle. Both are now known to be caused by an agent called a protein prion and hence transmissible via ingestion of brain tissue, but for some fortunate reason, scrapie is not passed on to man.

The rabbits we kept were used to some extent for food and I must have eaten it at some stage, but as I have knowingly eaten it only once since and it was not to my taste, it seems unlikely that I was given it often. There was bread and dripping, which is surprisingly good, although mention of it creates visible signs of distaste when young people today are told what it was. Grandma also made gorgeous rice puddings with a thick skin on top, covered in lots of grated nutmeg.

Our next-door neighbours were Sid and Nora and their two teenage children. Sid was a dark, rugged, taciturn man who worked on one of the docks. He had at some stage been in prison for stealing from the dock, and although he engaged little in conversation, he was always very civil to me and I liked him. I liked Nora too. She was friendly, but unfortunately, her personal hygiene left a lot to be desired and she smelt. The house stank. Fortunately, we rarely ventured in, only when asked. Those were the days when if someone ran out of something, such as sugar or butter, one would 'borrow' a cupful from a neighbour, either because shop opening hours were more limited or they had run out of money—and Nora often 'borrowed'. The poor woman had dreadful bilateral *talipes equinovarus* (club feet), and she thumped along with the outer sides of her feet flat on the ground. She had long, lank, unwashed hair and was edentulous but never wore false teeth. Nora always called Grandma 'Barmby' to her face, as if she was in the forces.

Next to Sid and Nora were the Priestmans, of whom my only memory is watching Mr P being carried out to an ambulance on a stretcher covered in scarlet blankets and, out of curiosity, approaching for a better view, only for my mother to drag me back, warning that I might catch what he had. People used to talk about the fever ambulance as if it was a different vehicle from the usual, and also about the Fever Hospital. There was a superstition that if you saw an ambulance you had to touch your collar and could not release your grip until you saw a four-legged animal, otherwise you might become ill yourself. Clearly, there was still a fear of infectious diseases. There was another woman who I remember well. Known as Gypsy Doris, middle-aged or elderly even then, with long, jet-black hair contrasting with bright red lipstick, my generation looked upon her with curiosity and even a little fear because of her witch-like appearance. I was an adult before I realised she was a prostitute, who apparently continued in her trade until she was in her eighties.

My mother received a war widow's pension but I suppose she wanted to supplement her income to pay for our board and lodging and so she went to work as a shop assistant at Robinson's Bakery, and remained there for the rest of her working life. The founder, Herbert Robinson, lived in the house that was an integral part of the shop. He was now retired, and the business was run by his two eldest sons, Len and Wilf. In the centre of the bakehouse, which was behind the shop, was a very large oven fed by a pyramidal stack of coke. It was always kept heated because it was uneconomical to light it up intermittently. After Herbert's death when I was eight, the adjoining house was offered to my mother. It suited her because she could at last be independent of her parents, did not need to travel to work and they deducted the rent from her wages. But it also tied her to the job. The shop opened at eight-thirty in the morning and she never got in before six in the evening.

My father's parents, Charles Clarence and Margaret Ann Clarke (nee Ridsdill), were both born in 1893. Dad was the second of their three sons. They lived at 7 Selina's Crescent, Rosmead Street, and I only saw them every few weeks. When we left Ivy Terrace I was old enough to go on my own. I regret to say that I regarded a visit as more of a duty than a pleasure, but they were always pleased to see me and gave me pocket money. I usually went on a Sunday and, especially when it was fine weather and windows were open, I could smell roast beef cooking throughout the whole journey, accompanied by *Forces' Favourites* on everyone's radio, with Jean

Metcalfe and Cliff Michelmore, and the familiar requests from BFN Osnabrück or other military bases. Granddad would sit to one side of the fire, smoking his roll-ups and completely dominating the conversation. He had an opinion about most topics (my mother used to say, somewhat disparagingly, that you could not tell him anything). But I always thought he talked a great deal of sense and that he was better educated than his occupation of flour miller would suggest. He once said that if you saw what went into white bread you would never eat it again. He told me York was 'the city of the three Rs'—if you can't get a job on the railway or at Rowntree's, you throw yourself into the river. And the cure for rheumatism was a pint of railwaymen's sweat. It took me a few years to understand that one!

Grandma sat quietly, saying very little and just smiling occasionally. I can't recall either of them telling me about my dad. I saw nothing unusual about this. After all, I'd never seen him, and as a boy was more interested in other things. As I grew older, I assumed it was to avoid painful memories for them, and to avoid highlighting to me the fact that I was growing up without a father. However, I much later discovered that Grandma's own father, John Henry Ridsdill, was a trawler skipper who served as a Royal Naval Reserve skipper during World War One, when his boat was converted to a minesweeper and became *HM Trawler Strymon*. The ship was sunk on October 27th 1917, when he was aged fifty. His body was swept ashore at Spurn Point. Grandma's elder brother, who was a first engineer, was killed on June 1st 1918 when a German torpedo hit his steam trawler, *Egret*. He was twenty-seven and married with a son. Poor Grandma Clarke, sitting quietly and smiling benignly, had lost her father, elder brother and then her middle son. All casualties of war.

Chapter Two

The Price of Fish

My first teacher at Mersey Street School was Miss Pinder, a kind, slim, youngish lady, with dark hair taken up in a bun, a flattened, wedge-shaped nose and a prominent gap between her upper front teeth. We wrote on slates with chalk and chanted our times tables in that monotonous rhythm of schoolchildren everywhere. There was only one male teacher, an old man known as Daddy Watson, about whom there was a song to the tune of Snow White's *Seven Dwarfs'* theme—

'Heigh-ho, heigh-ho, Daddy Watson's on the po, he's singing songs and dropping bombs, heigh-ho, heigh-ho, heigh-ho, heigh-ho.'

Rude lyrics have a special place in the memory banks.

Mersey Street school was only a short walk from home and was at the boundary between the streets of terraced houses off Holderness Road and the Garden Village. The latter was built by Sir James Reckitt before World War One, originally mainly for the firm's employees, and each street was named after the species of tree planted along it. Nevertheless, the pupils were essentially from the working class and, in fact, the school was known to others as "mucky Mersey". My mother told me never to go to the school lavatory for a number two because I might catch germs, whatever they were. Hence, despite all my efforts, I had an accident one afternoon during playtime and had to spend the last lesson squirming around while we sat in a circle listening to a story. The teacher never said a word, but she must have known. On a separate occasion, after several classmates had asked to go to the lavatory during a music lesson in the main hall, a different teacher lost her temper and threatened the next person who asked with a smack of the ruler, so naturally, I wet myself, leaving a nice pool on the parquet floor—and she went berserk.

One afternoon after school, Barry Lloyd and I went with Dougie Harrison to his house to play and at some point, Dougie found some of his dad's *Men Only* magazines. Barry later told his mam and the following day we were all hauled in turn to be questioned by the headmistress. It's an odd sensation, made to feel guilty for something you are supposed to have done wrong, but you aren't sure what. I'll bet Mr Harrison got it in the neck, though.

Despite being a war widow with a child, my mother was determined not to accept 'charity.' A teacher asked if anybody would like to go on a 'sunshine trip' to the seaside, probably Withernsea, Hull's equivalent of Blackpool. I put my hand up and was on the list, but when I told my mother she was adamant that I was not going because it was for poor children only.

The year 1953 was a remarkable one for Britain. The Brylcreem Boy, Denis Compton, scored the winning run in the final test against Australia, enabling England to regain the Ashes for the first time since 1932-33. At long last Gordon Richards won his first Derby and Stanley Matthews won an FA Cup winners' medal when Blackpool beat Bolton 4-3, having been 3-1 down with only twenty minutes to play. Neville Duke created a new world air speed record and, to cap it all, a British team climbed Everest just before the Coronation itself. Naturally, Coronation Day—June 2nd 1953—was a holiday, and we all went to Uncle Stan's to see it on his eight-inch Marconi television. Stan was Auntie Elsie's son, and although he was actually my cousin, he was so much older that he was always called 'Uncle'. Granddad stayed at home. He wasn't a republican—he had stayed up late to hear the announcement of George V's death in 1936—he just wanted some peace and quiet on his own. The procession and ceremony seemed to last all day, and we youngsters were bored by much of it.

We sat in pairs in class. My first companion was Anita Boothby, but she disappeared after a few weeks, having contracted tuberculosis. A pretty girl whose name I forget also left soon after spitting up blood—another case of TB. Ann Walton, the elder sister of my best friend, John Walton, was sent to Castle Hill Hospital, the local sanatorium for chest diseases, mainly TB, where she had a phrenic nerve crush procedure to rest the affected lung. The phrenic nerve, which supplies the diaphragm, arises in the neck and descends into the chest. It was thought that 'resting' a lung affected by the disease might aid natural recovery. They achieved this by crushing, but not dividing, the nerve via a small incision in the neck, the damaged nerve slowly recovering with time. Until it did so, the diaphragm on that side was paralysed, restricting movement of the lung. Drug treatment for TB was in its infancy then. Streptomycin was in use by 1946, although not widely, and isoniazid was not available until 1952.

My mother, influenced no doubt by my grandmother, was against any form of vaccination, so I did not receive the smallpox vaccine, nor any other that was available. Later on, at fourteen, when offered the test for prior exposure to TB by the Mantoux skin test, with a view to receive BCG vaccine if indicated, I was excluded.It was only when I became a medical student, and independent, that I was vaccinated against smallpox and polio. Because of my proximity to the two infected pupils and to Ann Walton, plus the many football and rugby league matches I attended, surrounded by people coughing and spitting all over the place, I was astounded to find that I was Mantoux negative. A small epidemic in 1961 prompted the smallpox vaccination and the microbiologist who attacked me made a thorough job of it, creating several puncture wounds. My arm swelled up like Popeye's biceps (only it was the deltoid—I know my anatomy) extending, without exaggerating, to mid-forearm. It was extremely painful, made agonisingly so by one of our year, Jean Price, a jolly-hockey-stick type of girl who could have been a prop forward. In response to a joshing remark of mine, she struck out hard and hit me directly over the sore site. Although the diphtheria vaccine that I did not have was developed in the 1920s, that for whooping cough was not available until 1949, well after I had already suffered from this nasty illness. My mother's abhorrence of vaccination did not extend, however, to the ancient ritual of circumcision that was inflicted on me at six weeks of age. Our GP, Dr Eddey, performed the procedure, ably assisted by the anaesthetist, Grandma Barmby, who apparently dropped chloroform onto a handkerchief!

They always judged me to be top of the class. I presume it was based on marks, although I can't recall any specific exams other than vague memories of the eleven-plus.

As Mother had no social life whatsoever, a trip into town on her Saturday afternoon off was a treat. She finished work at one o'clock and we then had a cooked lunch. Sometimes we waited for more than one bus because everyone else was doing the same thing, and they were often full.

I took for granted the extent to which the Germans had bombed the city centre. Whenever we called in the Edwin Davis department store, where Mother worked before she was married, she would look out for people she knew. One male assistant looked shifty. Although he always had a friendly conversation with Mother, the way he hunched his shoulders and looked sideways and avoided eye contact put me off. My attitude changed when she said he'd never been the same since the war. He was a prisoner of the Japanese and one of the lesser punishments he regularly received was having a metal bucket placed on his head, which was then beaten with a shovel.

I first played for our school soccer team when I was about eight and continued to do so until I left, aged eleven. Our strip was four chequers of blue and yellow and our coach was Mr Dickinson, a Geordie, who was therefore deemed knowledgeable about the game, Newcastle United being in the First Division and 'always' winning the FA Cup. He was a very cheerful, handsome man, with wavy black, Brylcreemed hair and a large dimple in his chin. My first game was at left full-back on a freezing day in continuous rain. As we won 10-1, I had little to do. Poor old Granddad came to see me and stood in the rain for the entire game, the only spectator apart from the teachers. When I got home, I turned on the Ascot water heater to warm my frozen hands and had never experienced pain like it as the circulation returned.

Although I was still keen on rugby league, as I was now in the soccer team, I started watching Hull City at Boothferry Park. For as long as I knew them, City were in the Third Division (North). There was also a Third Division (South) and I assume the geographical separation was to reduce the travelling and subsistence costs in those days of huge attendances but cheap admission prices and no TV revenue. Men often lifted us down to the front if there was a big crowd. We were always behind one or other goal, probably the cheapest stands, and, of course, everyone did stand. Almost every man wore a flat cap and nearly everybody smoked.

One evening, City were playing Wolverhampton Wanderers of the first division in a friendly match. I had inadvertently left my football boots on the bus after an away match a day or two earlier, and together with a friend, had just retrieved them from the Lost Property office (can you imagine that happening now?). With our autograph books, we went to the nearby Station Hotel, where the visiting players were staying, not expecting to get in—small boys did not enter posh hotels unaccompanied. But the top-hatted, uniformed commissioner kindly ushered us into the coffee lounge just within the entrance. There, seated on sofas, was the Wolves team. I immediately recognised the goalie, Bert Williams, blond and handsome, because he featured in my current edition of *Stanley Matthews' Football Album*. But where was the Wolves and England captain, Billy Wright? Oh! What a relief—he had just entered. I approached him and politely asked for his signature, whereupon he smiled, obliged, and then, in his Shropshire accent, said, 'You've got ink all over your face, son.' (My Biro had been leaking, unknown to me). Much to my disappointment, Wolves lost the game 3-1. I like to think they didn't try too hard—one's heroes must not be seen to be fallible. Do I still have that precious page? No, I lost the autograph book on Anlaby Road Cricket Ground a few years later while in the vain hope of getting Fred Trueman to sign.

Hull Fair took place every October on spare ground in Walton Street. Political correctness and concerns for animal welfare have decimated some of the attractions with which we were familiar. There were boxing booths in which a member of the public could earn five pounds if he could survive three two-minute rounds with one of the professionals. Every now and then some chap was successful, probably allowed to succeed to maintain interest. Most of those who came forward looked as if they really needed the money, but if some tough guy who fancied himself got up, the professional soon showed him the door. There were the usual freaks, no longer on show, fortunately—except, of course, on Channel 5—and a tattooed lady is no longer a novelty. I liked the old-fashioned roundabouts with the horses that went up and down, accompanied by the steam-driven organ. When we were teenagers our favourite was the Caterpillar, a roundabout with compartments for two-to-four people, which gently undulated and after a while, a large, green canvas cover would swing over so you were in darkness. If you had arranged to sit next to your girlfriend, you could steal a kiss. You had to win, rather than buy, a coconut by knocking one off a shy or firing an airgun or darts. Candy floss, cinder toffee and toffee apples were also a must, and a bit of an overrated treat, unique to the fair then, was attacking pomegranate seeds with a pin. There was a helter-skelter, ghost train, the wall of death and, of course, the waltzers, made more thrilling by the rough lads manning the ride spinning the cars as they passed. It was all noisy, crowded and exciting.

We were still in the age of the steam engine and trainspotting was a popular pastime. For the price of a bus fare into town and a 2d platform ticket one could spend most of a day out of trouble. Unfortunately for us, Hull Paragon railway station was a dead-end—you went to Hull because you were going to Hull, not through it on the way to somewhere else—so we had soon run out of engines to record. The solution was to go to York, on the main London-Edinburgh line, during the school holidays. There we saw passenger trains, not just shunters or dull freight trains with black engines. There were green and chocolate-brown ones, engines with 'blinkers' and plenty of 'namers'. Once in a blue moon we saw a 'streak', the nickname for the curved, streamlined Nigel Gresley-designed A4s, of which there were thirty-five. On one occasion I heard a tremendous rumble and hiss of steam behind my back and turned to see Mallard, A4, 4468, slowly but majestically make her way through. This was undoubtedly my most thrilling trainspotting experience, seeing the holder of the world record speed for a steam engine of 126 miles per hour in action.

Hull's Old Town was an interesting and compact area to explore. As well as the pier, where one could observe shipping and the Humber Ferry paddle-steamer, the Princes and Humber docks were still active and not out of bounds, even to children, even though there was nothing to stop you falling in. Nearby, on Humber Street, were bustling warehouses for imported fruit and vegetables and on one side of Holy Trinity was a covered meat and green market. Opposite this was the weekly general market with stalls of clothes and miscellaneous items. We enjoyed watching the traders selling sets of crockery with their rapid patter, bringing the price down, as in a Dutch auction.

'Not a pound, madam, not even fifteen shillings, not to you, my dear, ten shillings—seven and six and it's yours!'

Opposite Holy Trinity in the Market Place is a gilded statue of King William III (Good King Billy to the locals) astride a horse. The city closed its gates to Charles I in one of the first acts of the Civil War and people regarded King Billy as the great deliverer after 1688. Granddad kidded me that when the clock struck twelve, the king would dismount and cross the road to the pub.

On Christmas morning I would creep downstairs to find Granddad already in his chair, pipe lit. Using it as a pointer, he would indicate a half-empty glass on the table, with a single segment removed from a cake.

'He's been,' he would say.

At the end of Christmas Day, Granddad gave a sigh, puffed his pipe and declared, 'Well, that's it for another year.'

The most famous act I saw at the New Theatre was George Formby, a favourite of my grandparents. Grandma, Mother and I saw him in the late 1950s. By then in his fifties, he had put on weight compared to his appearance in his prime, when he was earning enormous sums from films. At the end of his act he cheerfully introduced from the wings his wife and manager, Beryl, a former champion clog-dancer. I enjoyed his famous amusing songs and his unique method of playing the banjolele, but the *double entendres* were lost on me.

There was food rationing during the war and for several years afterwards, but fish was not rationed. Hull had many fish and chip shops, four of which were close to our house. Even so, it was only an occasional treat when I was younger.

Hull was the principal distant fishing port in Britain by the early 20th century and St Andrew's Dock in West Hull was the largest fish dock in the world. Trawlers were at sea for three weeks, catching 100 to 200 tons of fish, mostly cod and haddock. While their husbands were at sea, the trawlermen's wives received a weekly allowance from the owner for housekeeping expenses. The men were home only for forty-eight to seventy-two hours at a time and were known as the 'Three-Day Millionaires' because they would, understandably, blow a lot of their earnings on entertainment (ie drink). I often saw trawlermen on a binge and could spot them a mile off. They had their own uniform—made-to-measure suits in a special style, horizontal pleats in the back of the jacket and bell-bottom trousers as worn in the Royal Navy. But it was the colours that were most distinctive. In those days, most men's suits were pretty drab—light or dark grey, possibly brown. But those of the fishermen were sky blue, emerald green or very light brown. Invariably accompanied by a woman (or two), they had a taxi with them for the duration of their expedition, to ferry them from pub to pub. They were known to be very generous with presents for the children. Being a trawlerman was rewarding financially, but it was hard, cold, dangerous work and there was no guarantee of a good catch. People said you had to sail with a lucky skipper. I recall several trawlers going down with all hands, sometimes not because of storms, but capsizing by the sheer weight of ice accumulating on the superstructure. This vast industry was in the end destroyed rather quickly by the several Cod Wars, as Iceland was allowed to progressively extend its fishing limits. Its loss has been a grievous blow to the city of Hull. Along with the 3,500 trawlermen thrown out of work, those onshore who were dependent on their existence were also left without employment. But, as with the blitz, nobody else seemed to care.

Occasionally, in the summer months, my mother and I would go on a so-called Mystery Trip run by East Yorkshire Motor Services. After experiencing a few, there was little mystery. It was either east to Paull, a small village on the Humber estuary, or west to South Cave. At some point there would be a stop at a pub for the adults, the children being left outside with a lemonade and a packet of crisps. But mother never went into the pub. A certain sadness descends over me whenever I think of these excursions. They were all my mother had to break the monotony and she had no social contact with anyone of her age. In many ways, my presence ruined her life. There were thousands in her situation, but she was the only one known to me.

One day I was chasing a boy in the playground after school and just as I was catching him, he swung the high metal gate behind him and it struck me on the left forearm. It was pretty painful but wore off after a few days. About a week later, I felt comfortable enough to participate in a game of British Bulldog. Using one's folded arms as a battering ram, you approached an opponent while hopping on one leg and tried to unbalance them. The loser was the first to put his other foot on the ground. This boy and I collided and I felt a sickening pain in the left arm and couldn't continue. I noticed the arm was now bent, but said nothing. That Sunday I was at Grandma's with my mother and was asked to try on a shirt. As I winced and contorted my upper body to protect the arm, Grandma noticed something wrong and they whisked me off to Victoria Children's Hospital where I was examined by a doctor, who I think may have been the first black man I had seen in the flesh, and a cast was applied. I had sustained a greenstick fracture of the ulna in the first instance, with little or no deformity, but bent by the collision when playing British Bulldogs. The following day, after a cooked lunch, they took me to the Fracture Clinic at Hull Royal Infirmary, where the sister administered an anaesthetic so they could manipulate my arm. Like most children, I didn't like the smell of the mask, nor the sense of oppression it produced.

'I don't want gas,' I said.

'We don't always get what we want in this life,' she said, clamping the mask on my face.

That's the last I remember until I woke up with a new plaster not just on my arm, but over much of the rest of me.

'By, you can fight young man,' I heard a half-amused Mr Tatham, the consultant, say, before I turned away to vomit my lunch into a kidney dish produced
like a magician's rabbit from a hat by the sister. I was also sick in the taxi back
home. I should have been told to not eat or drink for four-to-six hours before the
appointment and should not have had a general anaesthetic on a full stomach. It
was fortunate, for me at any rate, that I vomited rather than inhaled my stomach
contents, otherwise I might not be writing this. A combination of nitrous oxide
and oxygen cannot provide deep anaesthesia, only enough to do a quick job, such
as a tooth extraction, or, in my case, an orthopaedic manipulation, 'bone-setting'.
My laryngeal reflex would probably have remained intact. Nevertheless, it was an
unnecessarily unpleasant experience for all concerned.

After six weeks my plaster was removed and Granddad gave me a sweet on the
bus returning from the hospital. The moment I bit into it I felt a tooth crack,
followed immediately by toothache. We had never liked the rough-and-ready
approach of the NHS clinic in Morritt Street. Even as a small boy, sitting back
in the chair, looking up at the dentist I thought to myself, 'If he's a dentist, why
are his own teeth so bad?' Instead, my mother contacted a private one, Mr W
Stamford Brittain. He was a heavily built man, getting on in years, and he had a
grey moustache. He was assisted by his equally large wife, who had an excess of
facial hair, around the chin as well as the upper lip. I do not mean this comment
unkindly; they were very nice people, a childless couple devoted to one another,
and always very pleasant to me during the ten years or so that I was a patient.
When I saw him for the first time after starting medical school, Mr Brittain asked
me how I had coped with anatomical dissection.

'Fine,' I replied, telling him we had got used to it quicker than I expected.

He said he had had no difficulty either, adding, 'I saw enough on the fields of
Flanders to do me for a lifetime.'

I was deeply saddened to hear that after his wife's death from cancer, the
poor man was so desolate without her that he took a shotgun to himself in the
bathroom. He attached a note to the bathroom door, requesting no-one to enter,
but to send for his best friend, who was named on the piece of paper.

Ahead of the eleven-plus, parents had to choose a preferred school in the event of their offspring satisfying the examiners. The headteacher, Mr Canham, suggested that my mother put Hymers College as her first choice. Hymers was a direct grant mainly independent school, regarded as the best boys' school in Hull in terms of its academic results, and only one other boy from Mersey Street had ever won a place there. Mr Canham thought I had a chance of being the second and he also planned to enter me for a governors' scholarship.

I hardly remember the eleven-plus exam at all. I was not aware of having done well and felt strangely neutral, although relieved at least that it wasn't a disaster. The governors' scholarship was a whole day at a later date. Again, I have little recollection of the exam, except two questions in the general knowledge paper. In answer to 'What is (or was) El Alamein?' I wrote 'The Sheik of Araby' and to 'Name a resident British bird', instead of being sensible and answering house sparrow or blackbird, because of my knowledge of ornithology I answered, 'Dartford warbler'. While that is correct, I may have been marked down because the examiner might either think, reasonably, that all warblers are migrants, unless he could be bothered to look it up, or more likely, that we don't want this cocky little blighter here.

But I do remember seeing the school—I loved it at first sight. Apart from the beautiful structure, there were twenty-five acres of playing fields surrounded by woodland. At lunchtime, in front of the main building, young men, not boys, in whites, with proper pads and gloves and real bats, were practising in cricket nets with a red leather ball. All boys wore grey, with red-and-black striped ties, and the same colours on their stocking tops, and some also had the red blazers of summer term. It was like something out of Billy Bunter's Greyfriars School or Jennings' Linbury Court in the Anthony Buckeridge novels. And there was a tuck shop! How I wanted to get in.

***

In May of that year, 1954, Granddad became ill. At first I was not aware that his illness was serious. But as the days passed and he remained upstairs, and with Mother being not reassuring, non-committal and looking upset in response to my queries, I began to feel scared. Although he had contracted pneumonia at least twice in the past, he seemed to have robust health. I visited him alone on his 77th birthday, May 11th, and took a box of handkerchiefs as a present. He thanked me and gave me a kiss. I mentioned Hymers College.

'Ah, the snobs' school!' he said.

Then, from his bed upstairs, he stared across at the slate roofs of Endymion Street opposite.

'Whose is that dog running across the roof?' he asked.

I was bewildered, not understanding that he was delirious and having a hallucination. It all seemed unreal. I knew he was ill and just wanted it to go away and not to have to think about it. On the night of May 13th we went to see him. Grandma, Aunt Elsie and Aunt Flo were there as well. He was in a striped nightshirt, confused and restless.

The next morning I got up and went downstairs, meeting Mother as she came in the front door.

'Is Granddad any better?' I said.

She tried to control herself, looking only a little upset, but then sobbed and told me he had died. I can't recall my reaction, but she eventually asked if I would like to see him.

'Yes, for a last time,' I said, and followed her out.

On entering the house, Grandma was in her usual chair.

'You'll look after me now, won't you?' she said when she saw me—a strange thing to say to an eleven-year-old.

We went upstairs where Auntie Flo gently said, 'He's just asleep, David, touch his hands.'

They were still warm, but I didn't like the muslin encircling his head and chin and tied above, to stop his mouth from gaping open. He resembled that picture of

Marley's ghost in my copy of *A Christmas Carol*, but at least he looked tranquil. I remember a sickening feeling of a terrible emptiness. And then I went to school.

I didn't go to the funeral. Mother 'protecting' me as usual. But I was told that as the cortege went up Barnsley Street there were men standing in tribute and doffing their caps.

Granddad died on the same date as my dad, May 14th. He tried to replace the father I never knew, and I loved him for it, and as I grew older, I appreciated more and more what he did for me. This man was one of twelve children, with a basic Victorian education. He was employed first as a farm labourer and then as a council worker on a low wage, bringing up three daughters in poor, cramped rented accommodation with primitive sanitary facilities and no hot water. Granddad had virtually no holidays to speak of and never went abroad. He lived through the trauma of two world wars, losing one brother-in-law and two sons-in-law. And all the time, he retained the dignity and kindness of the true gentleman he was.

Chapter Three

The Snobs' School

Mr Canham greeted me with my entry exam result as I walked in one morning. I had got into the snobs' school. I was delighted, but also embarrassed when he asked for applause for me at assembly. When I went home for dinner, my mother didn't realise I already knew the result, and from fifty yards away I could see her waving the letter furiously.

The uniform was a red cap with black peak and a badge with the coat of arms—a white hind on a green mound with a coronet around its neck, superimposed on the three crowns of the city of Hull, and a grey blazer with the badge on the pocket. The red blazer of summer with the black edging was optional but, of course, she bought it. Short trousers for now, leather satchel, black and red rugby shirts. Cricket whites could wait. She was so proud that she made me wear the outfit before I had set foot in the school.

Early September 1954 and the big day. Two bus journeys to cover the three miles. Three halfpence ('Three 'apenny, please'—if you pronounced the aitch everybody turned round to look at you, posh boy). So this was my new school. What a contrast to Mersey Street Juniors! The grounds were large enough for five rugby pitches and facilities included a pavilion and changing room, squash and fives courts, a shooting range and cinder running track. The headmaster, Harry Robert Roach, lived on the site with his family and his faithful labrador, Dinah.

We were assigned to our forms and, to my dismay and not a little surprise, I was in Lower 3B, with Mr Lund ('Sally', a pun on the cake, Sally Lunn) as form master. I realised that the privately paid pupils had benefited from a better education so far and that with effort I could get promoted to the A-stream.

The staff were all-male and most were Oxbridge graduates, a remarkable fact. Form masters sat with their pupils at assembly. When everyone was quietly seated, the porter left his sanctuary and walked the ten paces or so to the headmaster's study, knocked and put his head round the door to announce that all was ready. As the head emerged, the entire school stood until he mounted the rostrum, gave a slight bow, instructed us to sit and a prefect rose to read the lesson.

Harry Roach had been headmaster since 1951 and was only the third headmaster in the fifty-eight years of the school's existence. I always thought he was strict, but apparently he was benign compared to his predecessor, WV Cavill, and I cannot recall him caning anyone.

Two incidents from assembly come to mind. At its conclusion one day close to the end of term during my first year, Harry made an announcement and I can remember what he said almost word for word.

'I have suspended Burton and Barkworth of Upper 3A for wearing duffle coats. They tell me that it does not say in the school rules that one must not wear a duffle coat. That is quite correct. Nor does it say in the school rules that one must not come down Hymers Avenue on roller skates wearing a jockstrap and a fez.'

Almost as one, the school rocked forward to contain the laughter that we dare not release audibly, as the mental image of his description imprinted itself on our brains. Their parents withdrew both boys from the school. Seven years later when I registered at Medical School, I recognised this lad whose jet-black hair, dark complexion like a permanent suntan and bright blue eyes were unforgettable. I approached him and asked if he was Eric Barkworth. He looked amazed, admitted that he was, and burst out laughing when I explained. Being suspended, of course, he hadn't heard the immortal line.

The other episode was when an internationally famous singer, Alfred Deller, was appearing in Hull. One of the masters, probably Mr Watson, the music master, knew him socially. The headmaster introduced him as one of the world's greatest countertenors and we were told he would sing for us. He was a small man with a beard. Following a brief introduction on the piano, he opened his mouth and out came the voice of a ten-year-old boy. Once again, boys were biting their cheeks and covering their faces as they rocked forward in silent hilarity. None of us knew that a countertenor had the voice of a castrato, or boy treble, and the sight of a grown man, albeit a small one, especially with the compensatory beard, possessing such a voice came as a complete surprise. I only hope the poor chap failed to notice our attempts to conceal our amusement.

Wednesday afternoon was a half-day, ostensibly for sport, which meant rugby union. I avoided playing rugby at all costs, unless I was compelled to do so. Although I was keen on watching rugby league, I was skinny and did not relish physical contact. At least in league, once you were tackled you were released to play the ball, whereas in union it seemed a free-for-all. Furthermore, the games master, Bill Minns, never properly explained the rules, and seemed to favour the big lads who were already keen on the game. The only time I scored a try he disallowed it and I still don't know why. There was a boy in my class, McAusland, who eventually became captain of the First XV and was built like a butcher's dog. His father had been a keen player and was an old boy of the school, so he inculcated him in the game. In one compulsory game, he and I both dived for the ball. Naturally, his bullet-like head struck mine on the left side of the orbit, splitting it wide open, which meant a trip to the infirmary for stitches. The school arranged for a taxi but did not inform my mother, and when I arrived home well after six o'clock she was worried sick. I have never accepted that such team games are character-building. In any case, they are not strictly team games. The only ones who pass are those, like me, who wanted to get rid of the ball. The McAuslands of this world put their heads down and charge forward, daring anyone to go for the bulky piston-like legs and get kneed in the mouth or struck, like me, by their genetically determined thick skull.

I settled down quickly in Lower 3B. I made friends and was happy. The only subject I found difficult was maths. I had always been good at arithmetic, but mathematics is different. I was annoyed at this. I knew maths was a logical subject and that my inability was probably due to a basic lack of intelligence, and that what ability I had might be due to a good memory. I've always had difficulty thinking in the abstract.

Each class had twenty-nine pupils—why it was always this number I don't know. I enjoyed Latin and was good at the subject, as well as at French and German, but I was more interested in the sciences. I was always somewhere in the twenties in maths but in the top five in almost every other subject, and at the end of the year I was promoted to Upper 3A.

There was no male member of my family to teach me the facts of life, and I'm sure my mother never thought of it. On the bus to school one day, two boys in front of me were secretively looking at a book. I leaned over to see why they were so furtive and called out in a loud voice, 'Ooh, a sperm!', reading the words beneath the drawing of a tadpole-like creature.

'Shut up, Clarke, you idiot,' was the reaction as they closed the book and tried to look innocent.

It dawned on me that I had been missing out on something, although I was still not sure what. There was no provision at school for such education and they did not teach biology until the sixth form.

In the summer of 1956, my mother and I went on holiday to London for a week. My mother worked with Beattie Gritten, a dark-haired lady with a slim figure, and a self-confessed good-time girl during the war. She met and married a sailor in the Royal Navy, Johnnie Gritten, and they had a son, Geoffrey, a couple of years younger than me. Johnnie was the son of a Conservative MP, and met Beattie when his ship was docked in Hull for maintenance. They married after a swift romance. Johnnie later asked her for a divorce, opining that the marriage was a mistake and he had met and wanted to marry a Yugoslavian concert pianist. Johnnie became a communist and was a journalist on the *Daily Worker*. The holiday was prompted by Johnnie offering to vacate his Westbourne Grove flat for a week and agreeing to my mother and I accompanying Beattie and Geoffrey. Mother and Beattie identified with one another as wartime brides who were single mothers, albeit under different circumstances. The flat was near the Royal Oak Station and I enjoyed becoming familiar with the famous Tube. Johnnie seemed sympathetic to me and my mother, and I now know why. He showed us the *Daily Worker* offices and the printing press. I did not know what a communist was, except that it was something not quite acceptable. After visiting Dr Johnson's house, we went for lunch in Ye Olde Cheshire Cheese, just off Fleet Street, which was noisy, smoky and packed with journalists. Fleet Street was still in its prime as the hotbed of journalism, and I was probably in the company of several household names.

We were taken to a Chinese restaurant in Soho, a first for us. I especially enjoyed the lychees, but on the walk back I doubled up in pain, which turned out to be wind as a consequence of my first exposure to Pepsi-Cola. We were also given tickets for Al Read's *Such is Life* show at the Adelphi Theatre in the Strand. The show was probably chosen for us because Al Read was a popular Lancastrian comedian whose radio show attracted up to thirty-five million—yes, million—listeners. He was known for his comic monologues, usually taking the form of his wife nagging him, and he did all the voices himself. His catchphrases were 'That's the wife from the kitchen' and 'Ave yer finished, 'ave yer done?' The highlight of the show was the appearance of a young singer who was taking the place of someone else. She was called Shirley Bassey. Although she was only eighteen, Shirley belted it out as only she could. The theatre wasn't full and to me, my mother's applause seemed to stand out—I've always said Mother 'made' Shirley Bassey. In the next evening's *Standard* there was a report to the effect that 'a star is born'.

Later in life I discovered that Johnnie Gritten had published a book called *Full Circle: Log of the Navy's No. 1 Conscript*, a fascinating account of how in June 1939, as a twenty-year-old *Daily Mail* journalist, he was among the first to be conscripted in peacetime to do six months' preliminary training in the Royal Navy. To achieve publicity he was made RN Special Reservist No. 1 and expected to write about his experience in his paper. Of course, war broke out, and he ended up serving over six years. He had an eventful war, serving on the destroyer *Afridi*, which was sunk by Ju 87 Stuka dive bombers in May 1940, with the loss of about 100 of the 350 men on board. He served as a boiler room stoker rather than an

officer and hence this account by an educated man with a literary ability must
be almost unique. After this traumatic episode he was sent to Hull to join the
Humber Boiler Cleaning Party. There he must have met and married Beattie.
He describes the horrific blitz on Hull and comments that he had not seen such
destruction until he visited Normandy following the D-Day bombardment. He
was the only official Naval reporter to beach on D-Day, after his landing craft
was holed. His descriptions of his shipmates, with their colourful language, black
humour and camaraderie, are vivid.

I understand Johnnie became disillusioned with Soviet communism after the
invasion of Hungary in 1956, not long after we met him, and I think he left
the *Daily Worker* then. He was indeed the son of a Conservative MP, Mr W
Howard Gritten, MP for 'The Hartlepools' (as it then was) for eighteen years.
He did marry a Yugoslav concert pianist and had three more sons, but other than
a faded reference to Beattie on his record card, neither she nor poor Geoffrey are
mentioned in the book.

***

It was in 4A that we joined the Combined Cadet Force (CCF). The concept
behind it was elitist—that in a future war, an 'officer' class would be primed
ready for rapid selection and further training. We met after school once weekly
and combined routine drill with map reading, artillery practice with a field gun,
shooting and signals. For routine evenings we simply wore a khaki belt over our
school clothes. It had to be kept Blancoed and the buckle Brassoed or we were
admonished, or even punished by some form of detention. Blanco and Brasso
were compounds used by generations of British soldiers to keep various parts of
their uniform in tip-top condition. They issued us with proper uniforms to wear
on inspections and field day. The NCOs were senior boys, usually sixth-formers
and also prefects. Giving boys a quasi-military authority to command showed
their true personalities better than their role as prefects did. They could be much
stricter and more serious and abrupt, any display of humour or even pleasure

being perceived as a sign of weakness. I should say, however, that in my seven years at the school I never encountered snobbery or significant bullying.

I enjoyed drill, although it could be boring, as any mistake had to be corrected by persistent repetition. One boy in my year, John Drewery, who was academically brilliant, could not learn to march. He insisted on flexing the straight arm with extension of the same leg so that he looked ridiculous. Everyone fell about laughing, to the NCOs' consternation. We learned to strip and clean our World War I Lee-Enfield rifles and occasionally a Bren gun. The boys in the artillery had the most boring job of cleaning the field gun, raising and lowering its barrel to dictate the range, inserting a wooden shell and getting a thrill by shouting 'Bang!'

We were taken to the shooting range, a concrete building at the furthest end of the playing fields, to shoot at targets using a .22 rifle. No-one showed us how to shoot, it was assumed to be obvious, the only advice given being to squeeze the trigger, not to pull it sharply. I ended up squinting through the wrong hole and scored twenty-five out of 100, one of the lowest scores (not helped by not yet realising that I needed glasses). The boots I was issued would have been suitable for Little Titch, the music hall comedian. Not only were they far too big, but they were also so supple that they were almost floppy. At the first opportunity my mother bought a properly fitting pair, an expense she could ill afford.

***

The field day was supposed to be the most exciting one of the year. The tradition was to fight a mock battle on Beverley Westwood, a beautiful area of undulating woodland and pasture near the racecourse that is common land, popular for picnics and lovers' trysts. As luck would have it, persistent heavy rain created a quagmire and Major Lund declared the frequent flights from nearby RAF Leconfield 'enemy aircraft', requiring us all to fall flat on the ground. We travelled back to school by bus afterwards and then had to get home, wringing wet through and covered from head to foot in mud. I had never regarded myself as a rebel, but when I thought of my poor mother, who worked hard enough as it was, having to buy me boots and then wash this encrusted uniform, coupled with what I regarded as a useless playing-at-soldiers exercise, I asked if she would write a letter to the headmaster requesting that I leave the CCF. Harry submitted to our wish without demur. I thought at first that his benign acceptance was because he knew my father had been killed in the war—I carried my form's wreath from the main hall to Memorial Hall on Remembrance Day—but I later learned that he himself was a pacifist and didn't approve of the CCF.

***

In September 1957, we entered Form VA. I remember vividly sitting next to Mike Higham in assembly on the first day of term. As we quietly waited for Harry to emerge from his study, he whispered to me, 'This is the year we do O-levels.'

'What's O-levels?' I asked.

'They decide whether you stay on or leave,' he replied.

Immediately, a pang of terror swept through me. I was happy here. I could stay for the rest of my life. What would I do if they forced me to leave? My heart raced; I could almost hear it thumping and my ears started buzzing. It was all I could do to control myself and prevent a faint. Nowadays, we would recognise this as a panic attack. It heralded a series of similar episodes subsequently triggered by a new stressful experience, such as a claustrophobic situation in the middle of a row in church, and I could provoke one by simply thinking about it. Oddly enough, I had no problem with amateur dramatics. These attacks recurred at intervals throughout my life but became less frequent with age. Fortunately, I survived our marriage ceremony, but it took a fair bit of self-control. Much later, my wife gave me the useful tip of pressing a Yale key into the palm of the hand until it hurt as a means of distraction, and it works.

Another alarming incident happened that day. At the start of every school year we got a new form master who distributed new books and also acted as a pastoral tutor. Doctor Donald Finlay was our man—a shortish but broad, balding chap with a hooked nose and a fearsome reputation. Our lockers were being changed and Doc Finn, as he was called, read out the new numbers. I was sitting with Duncan McAusland, the tough rugby player, in front of me. Doc was speaking when he suddenly stopped, looked angry, and asked McAusland what was on his wrist.

'My new locker number, sir,' he replied, cheerfully, 'so I won't forget.'

Doc was silent for a while and then asked him to come forward. Looking puzzled, he did as he was told and showed his wrist. After a long period of silence, Doc suddenly swung out at the lad, clattering him across the face with enough force to knock him sideways. I remember the whole class semi-rising from our seats, thinking he had gone mad. It is fortunate that McAusland was such a well-built lad. Not only did he recover quickly, but he did not retaliate, which would have been unprecedented but natural under the circumstances. There was a pause and then Doc's shoulders started to heave up and down and finally, he gave an embarrassed laugh.

'Please forgive me McAusland, I am so sorry, I don't know what came over me,' he said. 'You see, I was one of the first to enter Belsen concentration camp and it all came back to me.'

The pen marks on the wrist resembled the tattooed identification marks of the inmates. We all immediately understood and felt sorry for him. There were no repercussions that I know of.

Doc Finn was an excellent teacher and a nice man, but he played on his reputation for being as fierce as he looked. He was responsible for one of the most amusing entries in a school report that I have encountered.

'This boy will be difficult to forget,' it said. 'But the effort will be worthwhile.'

Can you imagine a schoolmaster writing that today?

***

In 1957 the Asian flu pandemic reached Britain. I was one of the last to catch it and at one point there were only eight of us out of twenty-nine left in the class. When I did succumb, it was the first and only time so far in life when I doubted whether I would ever get better. The complete absence of taste and the lethargy were debilitating, and the post-viral depression an odd sensation to experience in a previously fit and energetic teenager. Thankfully, my mother remained well.

***

We took ten GCE O-levels in the summer. I was terrified of failing elementary maths, which in my opinion was not as elementary as it should have been. Unfortunately, I compensated for this in my revision by doing none for the Latin set books (*Caesar's Gallic Wars* and Vergil) because I was always in the top five for Latin and arrogantly thought that I would cope. I had no problem with the Latin grammar paper, but questions on the set books tested knowledge of their narratives rather than just translation, and I floundered. As expected, I failed additional maths, but also Latin, much to the headmaster's surprise. We didn't receive grades, simply a pass or fail, nor did we attend school and jump up and scream for the TV cameras—we listed our subjects on a stamped, addressed postcard and waited. I re-sat and passed Latin in the December and was later told that I obtained the highest mark seen in a re-sit. I must have scored pretty highly in the others, despite us not receiving any figures personally, because I was one of six or eight invited into the headmaster's study and entered for an Oxbridge scholarship. He knew I wanted to take biology in the sixth form and must have assumed that I planned to read Medicine (which was not originally my intention), so he entered me for Gonville and Caius, Cambridge. This entailed completing an application form, enclosing five pounds, and waiting for the A-level results in two years. My mother looked taken aback when I mentioned the five-pound fee but said no more. In my ignorance, I didn't realise until later that it was about a week's wages for her.

***

It was about this time that I acquired my first girlfriend. I travelled to school by two bus journeys. On leaving school at 4pm, if we ran and caught the bus at the end of Hymers Avenue, we might make Paragon bus and railway station for the 4.26pm home. The reason for the rush was that we would usually find three or four girls from French Convent, a private institution run by nuns, sitting at the back. I don't know why their parents wasted their money because academically it was a poor school, but academic success may not have been their chief motive. One girl was a plump redhead called Victoria Shores, the only child of a master mariner and his wife. We 'went out' for a few weeks and I was sure I would marry her, obviously, in about six years or so, but it inevitably fizzled out and she took up with my best friend Mike Higham, who lived much closer to her.

The summer of 1958 was pleasant in other ways. Another friend, Peter Allison, lived in the Oval in Garden Village, and close by were two brothers, Ian and Geoff Harris, whose father was master of the *SS Umgeni*, a merchantman, and was at sea for prolonged periods before returning with generous gifts. This summer we were offered packs of 200 Pall Mall cigarettes (American) for small change. Sadly, but inevitably in those days, we had been introduced to smoking, and sat in the middle of the Oval, puffing away. The Harrises had a large, fat black Labrador called Bo'sun who always greeted me by trying to copulate with one of my legs, much to the hilarity of the seated audience. Ian's birthday coincided with his father's return from a voyage, and at his party there was an enormous quantity of what looked like a congealed mass of tiny blackberries, but which we discovered was caviar. His last port of call had been Archangel in Russia.

Buddy Holly was at the height of his fame, and *Rave On* was riding high in the charts. By now I owned a Dansette record player and played the few records I owned repeatedly to try to elucidate what I was missing. I did not appreciate most pop music of the day and wasn't an Elvis fan until I was much older, but I liked Buddy. After his untimely death not much later, in February 1959, most of Upper VC came into school wearing black armbands. Another friend, John Walton, was keen on Duane Eddy and listened to the deep bass repetition of *Peter Gunn* ceaselessly. He also taught me a trick, which I regret to this day—he showed me how to inhale cigarette smoke. He had watched me and said that I smoked like a girl! I soon appreciated the pleasure of nicotine and by the upper sixth we were smoking Capstan Full Strength.

I loved cricket and was quite good with both bat and ball. One afternoon I scored thirty-five before being caught on the boundary. It was a big score for a schoolboy and George Dixon, the umpire and maths master, put my name forward for the second XI without my knowledge, until they requested me to attend for net practice. Instead of being thrilled, I was appalled, partly because I was two years younger than the rest and didn't want to mix with the older boys, but mainly because we had to pay for transport to away matches and I didn't want to burden my mother. Stupidly, I tried to burn one of my hands on the hot fireguard at home so I would have an excuse to not play. But I couldn't tolerate the pain long enough to cause an injury and finally I simply volunteered that I didn't think I was good enough, which was accepted.

Entry for sports day was more or less compulsory. I couldn't sprint and was useless at field events, so I plumped for the mile. I still smoked, although not yet at home, and had done no training, but I still fancied myself as a budding Roger Bannister. There were six or eight of us and I knew one of them, Bonner, was a good runner, but I was slightly older and not intimidated. My plan was to stick with him and take him on the final straight, just like on TV. The crucial issue was whether I could keep up with him. All went according to plan and I couldn't believe the ease with which I stuck just behind Bonner. On the fourth lap, just after the final bend, I stormed past him with about fifty yards to go. Chuff Dennis was manning the finishing line and just as I reached him, he waved his arm.

'One more lap, one more lap!' he shouted.

Honestly, which stupid person creates a running track with five laps to the mile? Well, it finished me and I was second-last. I sat in the pavilion afterwards for about half an hour before the waves of nausea eventually subsided after the effort I had expended.

Beyond the running track was a railway line, and across that, Anlaby Road cricket ground. This now-defunct venue used to host a Yorkshire County match once or twice annually. We could sneak in free by crossing the line after school (out of bounds, of course) and catch the last couple of hours of play. Yorkshire were playing Hampshire and had started an innings not long before we arrived, but they had just lost a wicket and sent the great fast bowler Fred Trueman in as nightwatchman. I expected this to be thrilling because Fred had a reputation for swinging the bat. Unfortunately for us, he fulfilled his role and played forward defensive strokes to almost every ball, posing by remaining in the stroke position to exaggerate it, with one foot remaining behind the crease, and playing to the crowd. But there was silence and then after yet another similar stroke and pose, a voice shouted, 'Big 'ead!' At close of play, with Fred undefeated, the teams strolled to the pavilion at the far end and we rushed onto the pitch in a forlorn attempt to get Fred's autograph. Halfway along, a policeman interrupted Fred and had a word in his ear. Both then changed tack and went to inspect a small black car parked at the edge of the pitch. It belonged to Fred—someone had slashed all of its tyres. Fred knelt to inspect the damage, with a small crowd gathering. I seem to think this was the first time I heard the F-word. Certainly, it was the first time I had heard it said so vehemently and frequently in such a short time, and in the presence of the law. Needless to say, we failed to obtain his autograph.

Chapter Four

A Lucky Escape

In September 1958, I entered the sixth form for two years of physics, chemistry, and at last, biology, before A-levels. By some quirk of fate, Mugs Hunter and I were the only ones taking biology. Because there were only the two of us, Mugs and I joined with the year above, the upper sixth. The first topic of study was *Rana temporaria* (the common frog). Frogs were killed using cotton wool impregnated with chloroform in a covered glass jar and then pinned out on a board for dissection. But to demonstrate the ability of a decerebrate frog to react to noxious stimuli they were 'pithed', and we had to do this ourselves. We picked the poor frog up by the feet and struck its head against the edge of the lab bench to render it unconscious. We then inserted a large optical pin into the skull and jiggled around to destroy the upper brain. Nervous reflexes were still intact, however, and if one applied acetic acid (vinegar) to a foot, the latter would be withdrawn as if in defence. I sometimes wonder whether there exists a God in the form of a frog waiting to wreak vengeance on those of us who have performed this distasteful procedure so many times. Ironically, the frog is one of my favourite creatures. I distinctly recall feeling uncomfortable and blushing when the reproductive system was discussed and the word testis mentioned. I was surprised on looking about that none of my new colleagues reacted at all—I was sure the odd smirk or titter would ensue. I was approaching sixteen at the time!

Our biology master was Frank F McNaughton. He was always known as Dan and was an Edinburgh graduate who played rugby for Yorkshire. He was sandy-haired, freckled and ruggedly handsome with his broken nose. He never wore a gown and always walked with both hands in his pockets, with the flap of his sports jacket perched over his prominent, muscular backside. The biology lab was an afterthought, a single-storey prefab isolated in an L between the junior school and the main science block, with woodland on the exposed side. As well as forms and desks for teaching before the dais and blackboard, there were benches with gas supplies and sinks and display cabinets containing purchased dissections in alcohol or formalin of frog, rabbit, dogfish and others.

Mugs was a real character. He was a keen coarse fisherman and also shot. I never saw him socially because he lived too far away, but he would have been a good friend. He was mischievous and always had a quip. To help us with physics Mugs and I had a weekly maths lesson from George Dixon. In the short corridor near the room was a table, piled with science magazines. Mugs and I usually arrived first and often browsed through these until George appeared. On one occasion, Mugs and I found our way separately and I went straight to the room and sat down to wait. A few minutes later a flustered George appeared followed by a sheepish-looking Mugs, and the lesson proceeded as usual. When George left, Mugs creased up. Unusually, George had arrived first and stopped to peruse the magazines, as we normally did, with his back to anyone approaching. Mugs arrived, mistook George for me and reached between his legs to grab his testicles before realising his mistake. George was a good sort, and I suspect that when he recovered from the shock, he had a chuckle to himself.

I once asked Dan if we could form a Natural History Society. Somewhat to my surprise, he reacted with enthusiasm. Our first visit was to the Leven Canal, an idyllic spot for pond life and plants. There was also a visit to Beverley Westwood hospital, where a bacteriologist named Dr Barnard showed us around his lab and fascinated us by the way he handled Petri dishes containing growing colonies of dangerous bacteria. He casually uncovered them while simultaneously chain-smoking, occasionally placing his upturned cigarette tip-side down on the bench when he needed both hands free.

During the summer of 1959, we spent a week in a hostel at Hutton Buscel, near Scarborough, conducting fieldwork and learning about the relatively new science of ecology. Dan brought his wife and his son, Neil, who was also at the school. We caught crayfish in Forge Valley and Mugs felled a cock pheasant in flight with a lucky shot from a catapult. At night we went to country pubs, the landlords of which could not care less about our age. There were two interesting incidents. Firstly, one day at the beach while we were foraging in rock pools, Neil McNaughton went snorkelling and found an unexploded Second World War mine. Later we watched from the clifftop while the sappers exploded it, creating a thud and large plume of water. Then a local expert offered to show us a badger sett close to a new housing estate that was still being built. We went at dusk. We succeeded in viewing a family, a first for me. We were looking in a northerly direction when suddenly, apparently out of nowhere, appeared a bright light to the north-east, above the sea. It remained stationary for a few moments and then rapidly moved across the sky at the same height, well to the west. Then it stopped, remained still once again and disappeared. Everybody agreed it was a UFO, in the strictest sense, in that it was an unidentified flying object. When we returned home, one of the boys wrote to the *Hull Daily Mail*, which published his letter. Interestingly, the same description was given by an observer in Aberdeen, but the direction of travel was from left to right, indicating that whatever it was, the object was somewhere in-between. The mystery remains unsolved, to my knowledge.

***

In September 1959, I entered the upper sixth, the year of A-levels. Mugs Hunter had left, and I was now alone studying biology. Dan said there was no point teaching me and just gave me his notes to copy out, but he did check my dissections of the cranial nerves and circulatory system of the dogfish and, believe it or not, the cockroach. I loathed handling the dogfish, with its sandpapery skin and the stench of formaldehyde. At least I didn't have to dissect the abdomen, for the liver of this creature must be one of the most offensive organs known and it is also parasitised by numerous roundworms. Urgh!

It was a quirk of fate that I was the only one in my year doing biology because there were about a dozen in the one above and a similar number in the year below. Being the senior biologist, I became the lab boy and earned £1 a week for tidying up at the end of the day, cleaning the blackboards and preparing certain experiments. Included in the syllabus was the dissection of a mammal, usually a rat or a rabbit. Dan told me to gas the rabbits he had obtained for the year below and pin them out on boards. No specific instructions, no advice—use your initiative. I put them in cardboard boxes sealed with Sellotape apart from an aperture through which was inserted a tube leading from a gas tap, of which there were several, feeding Bunsen burners. I then opened all the windows and switched on the gas. The poor things soon started thumping their paws, so I beat a hasty retreat for fifteen minutes or so. Fortunately, no mishap occurred, and after several sessions the dissections were complete, the carcasses being now virtually bare, impregnated with formalin and fixed in position like so many St Andrew's crosses. It was now time to dispose of them. I buried them in woodland a short distance away. Unfortunately, it was February, and the hard ground necessitated somewhat shallow graves. Some time later, Dan asked what I had done with them.

'Ah, that explains it,' he said when I told him. 'The headmaster has made a complaint. Dinah has made him a present of one. She dropped it at his feet.'

Dinah was the head's faithful labrador. Fortunately for me, that was the last I heard of it.

***

I remember Easter Sunday 1960 very well. We were at evensong and sitting near the back as usual. Terry Hudson, known as 'Pud', came in late and breathless, shuffled into the middle of the row behind.

'Eddie Cochran dead, Gene Vincent serious, pass it on,' he whispered.

It was the 'pass it on' that still amuses me. It was a common expression before the age of Twitter and texting, used to convey some piece of gossip or scandal at school, or here, in church. Rock 'n' roll singer Eddie Cochran, he of *Summertime Blues* and *C'mon Everybody*, which had contributed to the happiness of youth, had died in a car crash while on tour in England. He was just twenty-one. The taxi driver and the other passengers, who included fellow singer Vincent and Cochran's girlfriend, all recovered from their injuries.

Despite the approach of A-levels, I played John of Gaunt in Mo Mitchell's production of *Richard II*. Richard was played by Alan Dossor, who was the main character in every school play I can recall. Aside from his ability on stage, he possessed confidence bordering on arrogance and was always relaxed. At one point I had to kiss his hand and with the profuse perspiration created by greasepaint and lights, several beads of sweat ran onto it. He looked down at me with disdain, turned to the audience with an expression of mock distaste, and flicked the sweat away, creating a small ripple of laughter, which clearly was his intention. Away from the stage, we had no rapport with him. He was the star. It was obvious that he would try to make a career on the stage and he took a degree in drama, English and philosophy at Bristol, followed by a postgraduate acting course at the Bristol Old Vic school. But it is as a director that he made his reputation, mainly at the Liverpool Everyman Theatre and Nottingham Playhouse. He helped popularise the plays of Alan Bleasdale and Willie Russell, and also influenced the acting career of Antony Sher.

On Saturday afternoons we continued to watch rugby league. Being from east Hull, Hull KR should have been our team, but demonstrating the fickleness of youth we tended to watch the west one, Hull, because at the time they were the more successful. The game was still semi-professional and the bulk of some of the forwards consisted mainly of adipose tissue rather than skeletal muscle. In the pack were the Drake twins, Bill and Jim—Bill, tall and slimmer in the second row, Jim, shorter and bulkier, and a prop. Like Fred Trueman, the Drake twins lived in Heworth, near York. There was an often repeated tale that Freddie laid out both of them in a pub, having been accused of being a 'puff' because he was drinking lager.

When we watched Hull play we stood behind one of the goalposts on Bunkers Hill, a terraced elevation of wooden slats separated by cinders. There are Bunkers Hills all over the country, possibly deriving from Bunker Hill in Boston, Massachusetts, the site of one of the first battles in the American War of Independence. One Saturday afternoon, who should appear at the foot of Bunkers Hill but Eddie Waring, who presented TV's rugby league round-up on match days before becoming nationally famous for rugby commentaries and *It's A Knockout*. There was a widely shared view locally that he was biased against Hull and it was rumoured that he was banned from the press box at the Boulevard. The latter seemed to be confirmed by his presence in the crowd. He was wearing a beige mackintosh and what we called a Robin Hood hat—a narrow trilby with a decorative feather sprouting out of one side. I had just finished an apple and more in fun than hooliganism, I gently tossed the core in his direction. Much to my surprise, it not only landed in his hat but by some fluke it stayed there. Fortunately, all he did was turn around and give a pleasant smile. I think he was used to gentle abuse, as the crowd had greeted him with a few jeers. These days I would probably have been hauled out of the ground and banned for life.

There's a moment in the popular late 1950s' film *Doctor In The House* when Grimsdyke, played by Kenneth More, suddenly announces with a start during a sleepy post-Sunday lunch hiatus, 'Is that the date? Good God, it's finals in five weeks!' A similar feeling suddenly struck me one warm, sunny Saturday afternoon in spring, just as I was setting out to see the crowning of Sylvia McKay, briefly a girlfriend of mine, as the May Queen at St Columba's. A sensation of panic went through me as I remonstrated with myself for allowing time to pass without doing any revision for A-levels until this late stage. So there was no more going out until they were over. As usual, I had gone along with the flow and not planned things properly. I had no particular worries about passing biology or chemistry, but physics was another matter.

That summer Peter Allison's parents hired a cottage near Scarborough and invited me to keep him company. He and I cycled there, Peter in his usual competitive way determined to arrive before his parents, sister and grandmother, who left by car at about lunchtime. He continually urged me on despite my heavy bike being stuck in top gear, compared to his lightweight racer. During the holiday the dreaded postcard arrived, indicating that I had indeed passed biology and chemistry, but failed physics and the general paper. The latter didn't matter. I tried not to let the news put a damper on what remained of the holiday, having always had a Micawberish attitude that something would turn up. So, in September 1960, I joined the depleted numbers of the third-year sixth, my final year at school. Having passed biology, I extended the syllabus by studying botany and zoology and scholarship-level chemistry. Cod Watton was no longer our physics master, much to my relief. Instead we had a new man, Mr Brown, who was not long out of university himself. He persistently urged me to sit scholarship-level physics, stating that by being entered for it on his recommendation I was guaranteed an A-level pass, but I didn't believe him. At one point in the practical, when I connected my Wheatstone bridge the wrong way round, I wished I had followed his advice, but I rescued the situation by sneaking a look at Bruce Cutsforth's—inadvertently, of course.

At some stage I heard from Caius College, Cambridge, that I no longer featured in their plans. They should have waited for my next results. At the time I was disappointed, but I knew I had let myself down and had no-one else to blame. One problem was that I was happy at school, content to go with the flow, not thinking too much about the future, and had nobody in the family to either advise or cajole me. Hymers had a careers master, Mr O'Dell, but I suspect it was thought that with most boys being middle class, external advice was largely unnecessary.

Fortunately for me, I had a saviour. One day, out of the blue, Dan Mc-Naughton said to me, 'Isn't it time you started applying for medical school, Clarke?'

'Oh, I'm not going to medical school, sir,' I replied. 'My mother says we can't afford it.'

She still believed that to become a doctor, you had to 'buy' your way into a practice.

'Who on earth told you that? Well, if you aren't going to medical school, what are you going to do?'

'Er, I thought of possibly doing zoology.'

I remember the next sentence so clearly because he reinforced it by coming down from the dais to my level, looking me in the eye and wagging his finger.

'Listen, Clarke, if you do zoology, you'll end up like me, teaching it. Get yourself to medical school.'

And with that, he bounded back up to his desk, lifted the lid, threw application forms for Edinburgh, Leeds, London and Newcastle at me, and told me to complete them as soon as possible.

Medicine had seemed an unrealistic ambition because of those financial concerns, plus the fact that it was a five-year course and I knew next to nothing about local authority grants. My mother seemed to think I might make a chemist, by which she meant a pharmacist. I think she had visions of me in a short white coat in Boots on Holderness Road.

I was interviewed at King's College Hospital on Denmark Hill, where they toyed with me, making me feel very small, and spent some time asking how I expected to be funded. I would never have forgiven them had I not been fortunate enough to get into a better medical school. Chris Day in the year above had gone to Newcastle where he was happy, so I applied too.

Arriving early for the interview, I strolled around the medical school building on a sunny day and memorised the motto '*Scire usum medendi*' ('To know the art of healing') on the coat of arms, just in case the interviewers tested my powers of observation and my Latin. But there were no trick questions. The two academics, who I never subsequently recognised, were smiley and pleasant and gave me the impression that my application would be successful should I pass physics. This impression was confirmed by post shortly after, and hence I made no other applications. It was a great relief. Firstly, I knew where I was going, and secondly, providing I passed the exams, my career was decided.

On looking through that well-known medical instrument, the retrospectro-scope, the two best things that happened to me were going to Hymers College and Newcastle Medical School. Before applying, I was unaware of the medical school in Newcastle. It seemed so much like second best, certainly to Cambridge, let alone universities such as London, Edinburgh or Leeds. But I soon found out that it was part of King's College, Newcastle, in the University of Durham, itself England's third-oldest university after Oxford and Cambridge, and founded in 1834. Had I succeeded in going to Caius College, one thing is for sure. I would not have had the great good fortune to meet and marry Mary Lack. Danny McNaughton with his finger-wagging warning against studying zoology—saved my bacon, and it is to my great regret that by the time I made the time and effort to convey my thanks, he had just died.

My last day at school was one I had been dreading. I had been there seven years and felt I could happily stay there forever—later on, I felt the same about being a medical student. The upper sixth occupied the balcony during assembly and I looked down on the main body of the school from a bench occupied by two others, between pillars. We sang hymn 317 from The Public School Hymn Book—*Lord, Dismiss Us With Thy Blessing*—which I had sung many times before. This time it was with some sadness, especially the last verse—'Let thy father-hand be shielding, All who here shall meet no more...' On this final occasion, the tears nearly came.

I took a labourer's job at Reckitt's, arranged by a foreman who was a customer in the shop. It was literally a wake-up call for me because work began at 7.30am, although fortunately Reckitt's was only a short bike ride away. I learned two important things in the first few days. Firstly, I was told by a foreman that students tended to work too hard and not to be guilty of this or, 'You'll have them all out on strike.' Secondly, one of my workmates was a certain Harry Chatterton, who had a dreadful limp, I presume post-poliomyelitis in origin, but who was well-known in Hull for his dance band. Harry was a delightful man whose piece of wise advice was, 'Always plead ignorance, David—I accidentally threw out £3,000 worth of equipment last week, but by pleading ignorance I got away with it.' Harry's band didn't provide him with enough income to leave Reckitt's.

The bulk of the workforce were female. I soon learned that the women did most of the work, while the men devised ways of avoiding it. The factory lasses were the salt of the earth, but they could be frightening. When I first started I was an item of curiosity, but they soon treated this skinny, bespectacled youth as a figure of fun rather than a prospective sexual partner. If any arrived late for work I had to take them up in the lift, and one day two girls came together. No sooner had I shut the gate and pressed the button than they screamed with laughter and made a dive for a certain part of my anatomy.

One day I had to accompany a lorry driver to the docks. He asked about me and said he knew my mother, as he sometimes came into the shop. Then, quite angrily, he said that as she was a widow, I should help her out by getting a job, never mind wanting to be a doctor. I could see what he meant, but it seemed very short-sighted.When the A-level results came, I was relieved to have passed the dreaded physics, together with all the others, and to have gained scholarship level in chemistry. Confirmation of my place at Newcastle followed, but because this occurred so late I was unable to obtain accommodation at a hall of residence and was offered lodgings with a Mrs Neylon of Roxburgh Terrace, Whitley Bay, on the coast and ten miles east of the city. My mother was just as concerned as I was and soon arranged with Mr Baines, a divorced commercial traveller and supplier of yeast to the bakery, to hire a car to visit the lodgings and ascertain whether they matched her exacting standards. My mother was happier than I was with the situation. The Neylons had a large Edwardian terraced house with a small, neat front garden. They were a smart, nicely spoken, old-fashioned couple, and they impressed Mother. So, accommodation accepted, we returned home.

Grandma kindly bought me a trunk for my possessions. I still have it, secreted away near the gas meter under the stairs.

I booked a place at the freshers' conference, held two days before official registration in October 1961. A cheque arrived for £100, my first term's grant out of £300 for the full year, the maximum available. The cheque was issued by the Midland Bank. I was under the impression that funds could only be accessed via this bank, so I opened an account at the Midland in Grainger Street. For the next few years I visited two bank clerks at this branch. The one with the shortest queue was Mr Strachan, who must have been approaching retirement. The one with the longest was a young, pretty, blonde girl called Kathy Secker, who later became a model and an anchor for Tyne Tees Television. I last saw her at the Town Moor Fair in the early 1980s. Despite the twenty-year interval, she had lost none of her beauty.

Chapter Five

The Cutting of the Cord

W ell, the day came for the parting—the second severing, so to speak, of the umbilical cord. My mother found a stand-in for the shop so she could see me off at Paragon Station. She was never one for displaying emotion. I don't know if that was inborn or acquired by the knocks she had taken. But as I kissed her goodbye, there was just a quiver of the chin and down-turned corners of the mouth as she realised that this was the beginning of the ultimate separation. Neither of us had shown much love for one another. Many times I felt unloved and lonely and she controlled my life fairly rigidly, but I'm sure she always thought it for the best. As I grew older, I realised that in many ways I'd been a burden to her—she was an attractive woman yet had no social life and little opportunity to meet a new man, with me as an encumbrance. She was only forty-five when I left, but was set in her ways.

After a three-and-a-half-hour journey by steam train, I arrived at Newcastle's Central Station and then took a bus along the Coast Road to Whitley Bay. The first thing I noticed as I took a seat near the conductor, who was in conversation with another passenger, was that I would have to learn a new language. In Hull, we had a hard 'u' and did not pronounce our aitches, but otherwise, I felt I spoke like a BBC announcer. These people gabbled at a rate that verged on the incomprehensible.

Together with my new roommate, Dick, a chemistry student from Redcar, I took the train into the city the following day for the start of the freshers' conference. We then walked to the King's College campus, conveniently situated in the city centre. The Crows Nest pub stood almost next to the short inclined street leading to the Students' Union, and then through an arch was a quiet square of raised lawn and gardens enclosed by university departments, including the Armstrong Building, where the meeting was held. Alongside this was the Medical School, which dated back to 1834, although the building I attended had opened in 1936. Both buildings were constructed of attractive, smooth, red brick, with the additional incorporation of stone decoration, as was the Royal Victoria Infirmary, or RVI as it was always known, opposite. It was one of the most beautiful teaching hospitals in the country. That first day I met a Geordie student who took me at lunchtime to the Brandling pub. It is a cliché, but true, that the new sensation of freedom in sinking a pint and smoking a cigarette without furtive looks of fear of detection was like a breath of fresh air.

A couple of days later we officially registered at the medical school and were given a brief welcome from the registrar himself, Norman Shott DFC, who was smooth and blond with an RAF accent.

'Welcome to the medical school,' he said pleasantly. 'I hope you will work hard and deserve your place because seventeen people have been turned down so that you can come here.'

These words certainly made an impression on me.

Within minutes I had an introduction to medical diagnosis. I heard a voice at the next table say, 'Are you feeling well?' and looked across to see an elegant West Indian lady, Margaret Hudson-Philips. She was addressing a new colleague, Nick Clapham. She had immediately recognised that he was faintly jaundiced. He had infectious hepatitis, and we didn't see him again for a month.

We were issued with a reading list and told to purchase a gown, which had to be worn for lectures and examinations. Most of my books were second hand, obtained from the porter's lodge as they were no longer required by the years above, but I have kept them all since. I did buy one new book—the famous *Gray's*

Anatomy, at six guineas (£6 6s), equivalent to about a week's upkeep. It was in fact, superfluous, because all our gross anatomy was learned from *Cunningham's Dissection Manuals*, which, purchased second-hand and used extensively in the dissecting room, were greasy from human fat, and smelled of embalming fluids phenol and formalin.

There were about ninety-five students in the year. Twenty-two were female. Eight came from overseas—two each from Nigeria and Ghana and one from India, Pakistan, Hong Kong and Norway. At least twenty-five had been educated at public or direct-grant schools. Ten or eleven men hailed from a single school, the Royal Grammar in Newcastle.

Officially, Medicine was a six-year course, but year one was known as the 1st MB and was reserved for those students who did not possess the requisite science A-levels. The 2nd MB was taken after eighteen months and included anatomy, physiology and biochemistry. Failures could take the examination again in three months and another failure meant dismissal. A significant number routinely failed and hence it was regarded as the major stumbling block on the way to graduation. As it approached there was genuine fear, but that seemed far away at the moment.

The frustrating aspect of the pre-clinical course was having no contact with genuine patients until we passed 2nd MB. However, new ideas were springing up in medical education, both in the UK and across the Atlantic, and a new integrated course was being planned. We were the last of the 'old curriculum', although the new multiple-choice questions were tested on us and compared with our performances in the standard essay questions. Oral examinations (*viva voces*, or simply *vivas*) played a major role throughout.

There was excitement mixed with apprehension at our initiation into anatomical dissection and meeting 'our' body. We crammed into a steeply tiered lecture theatre. On the podium was a table, upon which stood an upturned white enamel dish, guarded by a mortician bearing a suitably cadaverous grin. Dr Tom Barlow, senior lecturer in Anatomy, breezed in. He was a white-haired, ruddy-faced, cheerful Mancunian, smartly dressed in shirt and tie, with almost bell-bottomed trousers and a pristine white coat with a green collar. He turned the dish over to reveal a pallid human arm. Without further ado he showed us how to remove the skin and the sheet of connective tissue beneath it—the fascia—a term with which he seemed to think we were already familiar. Up we then trooped to the dissecting room on the top floor of the building, with frosted windows to prevent window cleaners from peering in and falling off their ladders. It contained about twenty metal tables in five rows of four, each with a cadaver covered with a transparent sheet under which damp cloths retained moisture. There was a strong smell of phenol. Six students were allocated to each body. We had our own instruments and, taking turns, conducted the whole dissection ourselves. Demonstrators supervised and periodically examined us with *vivas* on our progress. They were qualified doctors, having just completed their first compulsory hospital post (house jobs) and now embarking on a career in surgery. The first step towards this was passing the Primary FRCS (Fellow of the Royal College of Surgeons) and the best chance of achieving the qualification was by working as a demonstrator, as opposed to taking on a junior surgical post with its long working week and onerous on-call duty, in which serious study was almost impossible.

We spent too much time on anatomy—three hours, four afternoons a week. The first term was dedicated to dissecting the upper limb. In addition, there were lectures on embryology and neuroanatomy, with microscopic studies of both. These were given by the head of department, Professor Raymond J Scothorne, a tall, middle-aged man, with sandy hair and moustache, striking blue eyes and a slightly amused expression. His ability to create multi-coloured chalk illustrations of complicated embryological development on the blackboard in real-time was astounding. He could improvise, too. Once, he asked if everyone understood a certain process that he had just drawn. Some looked puzzled, so he thought for a moment, and then recreated the image from an entirely different angle, as if he was situated within the body itself. Amazing.

After the first term it was thorax, and abdomen for the second; third term, lower limb; fourth, head and neck. Head and neck was the most difficult—so much crammed into such a small space.

Anatomy, physiology and pathology had historically been the core subjects of medicine. Crudely speaking, if you knew the structure of the body and its parts (anatomy) and how each part functioned (physiology) and what could go wrong with it (pathology), then you could become a doctor. But a change was in the air. Curriculum committees under the direction of the Dean of Medicine were planning radical changes, including the relative time allotted to different subjects, which was bound to cause friction, as professors strove to protect their own specialty.

We were advised to purchase half a skeleton, for ten guineas (£10.50). The bones were not articulated, that is, they were not joined up. There was a whole skull, with the vault sawn off, so that the complicated interior could be examined, one side of limb bones, ribs, half a pelvis, but all the vertebrae. Foot and hand bones were in bags. Professor Scothorne gave a wry smile and told us not to enquire as to their source. They arrived in crude wooden boxes with German script on the sides. Although the bones belonged to adults, they were small and thin and clearly came from the Indian subcontinent. One of our group, Mukunda Dev Mukherjee, said that in his home city of Calcutta it was common to have to step over the bodies of poor people who had died in the streets of poverty and hunger. It seemed that some enterprising organisation was harvesting such victims, boiling them down and selling them to the affluent West. I still have my little man—or, at least, half of him; he sleeps in the garage. I sometimes take him out of his box and look at his skull, partly refreshing my memory of the anatomy but also, like the gravedigger with that of Yorick's in *Hamlet*, pondering the sort of life the poor man had.

The time spent in the dissecting room did have one benefit—it helped to socialise us. We got to know one another and learned to work as a team. It is remarkable how quickly we transformed from being apprehensive about the prospect of exposure to dead bodies to becoming blasé and treating the place almost like a social club.

A few weeks after we started, the chief mortician, Tony Bolam, asked if I wanted to earn ten shillings (50p) by helping him collect a body donated for dissection.

Tony was in late middle age and seemed like a figure from the past. He was short, red-faced, bald with white hair at the fringes and wore half-moon spectacles. He always wore a rounded starched white collar attached to a striped shirt, a waistcoat with a watch chain and, to top it all, spats. We drove off in his van but because the coffin encroached onto the passenger seat, I had to sit on it. Our destination was Ashington, twenty miles north of Newcastle. When we found the address it was a small, neat bungalow, and a grieving family were in the front room. The deceased had willed his body to 'medical science', but the widow had not expected him to be collected so soon, his death having only occurred that morning. She was quietly weeping into a handkerchief and said she would prefer the vicar to call first. Not very tactfully, Tony observed that there was little the vicar could do now (clearly, he did not want to return empty-handed). To top it all, he cautiously hinted that if we delayed too long the deceased would, 'How shall we put it? Begin to...' and he gave a little sniff. The two adult sons picked up on the signal.

'Come on, Mam,' one said. 'Have a last look, then these men can take him away.'

I could have fallen through the floor when Tony added that this young man accompanying him would benefit from her husband's generosity. I sat on the coffin once again on the return journey. Back at the medical school, I saw the basement for the one and only time. There were two or three large tanks with cooling pipes attached and on opening one I was confronted with a vision of what it must have been like for soldiers entering Belsen. There was a jumbled pile of white naked bodies, including that of a child with an enormous hydrocephalus (water on the brain). Tony instructed his assistant to begin the embalming process on our new guest, and I collected my ten shillings.

After the freshers' conference, three lads joined Dick and I at our lodgings. One of them, Les, a mechanical engineering student from Oldham, was a full-of-life extravert, warm-hearted and as Lancashire as hotpot. My favourite remark of his was when he returned after a weekend at home, having presumably had some success in Oldham, as opposed to his lack of such with Geordie girls.

'They might wear clogs and curlers, but at least they do a turn,' he declared.

At night when we overlapped in the bathroom before bedtime, he would think nothing of washing his backside with a facecloth. If you looked surprised he would say not to worry, in the morning when he washed his face he'd have forgotten he'd used it on his arse the night before.

***

I got the impression that the physiology lecturers regarded us as an encumbrance to be tolerated, as we were distracting them from their important research. The professor of physiology was Alfred Alexander Harper, who discovered the hormone pancreozymin in the 1940s. He was a bachelor, always appeared stern and humourless, and lectured in a monotone. At 9am precisely the lecture theatre doors were locked so he wouldn't be disturbed by latecomers. When he gave seminars to small groups, he smoked incessantly. When he coughed he didn't remove the cigarette from his mouth, so that with the deep inhalation which naturally follows a cough he would fill his lungs with more irritant, which only exacerbated the situation. I used to wonder whether this was the action of an intelligent man.

I met people who I previously thought only appeared in novels or films. One such was a student called John Austin Forbes-Proctor, or 'Forbes', as he was always known. He was short, bent, with florid auburn hair and a quiff, wore patterned waistcoats, a bow tie and tweeds and smoked a pipe. His hoarse voice reflected his tobacco intake. Every lunchtime he visited the same pub for a sandwich and a few halves, and the crowd he befriended were a motley crew, much older in years. Sometimes a volunteer was required during our physiology practicals. On one occasion, the experiment was to test the effect of prolonged over-breathing on the pH of the urine. The theory is that by blowing off carbon dioxide from the lungs (carbonic acid), the blood will become alkaline. To counteract this, the kidneys will retain hydrogen ion ($H+$) to level things out and the urine will become alkaline, with a pH above seven. This is a perfectly simple and innocuous experiment, but even then most of us felt embarrassed to have to produce some urine for testing (albeit, not in front of our colleagues). The lecture theatre being triangular in shape, most of us arrived early and crammed into the apex to avoid being called out to volunteer. Forbes arrived last, from the pub as usual, and he and two others were each given a beaker and told to go next door and produce a specimen. The most acid urine would be chosen so that the desired effect could be more clearly detected. Shortly afterwards all three returned, two with beakers half-full of amber liquid, creasing themselves with laughter, and Forbes gingerly teetering along with his beaker brimming with what resembled pure water. Forbes was chosen and laid on a couch, settled his hands across his chest and closed his eyes. Encouraged to breathe deeply and rapidly, he merely remained in repose with a tranquil expression on his face. The experiment ended in farce, but it was only when we departed that his companions could disclose that when they adjourned to the empty theatre next door, because of his liquid lunch, once Forbes filled his beaker he couldn't stop peeing, and emptied the rest of his bladder up the wall.

Another experiment which had amusing consequences, in the year above ours, was demonstrating the effect of emotion on blood flow and pressure. The volunteer placed his hand and forearm in a glass compartment sealed by a rubber flange around the upper arm. Beyond the hand end was a rubber cover attached to a lever and a rotating smoked drum. Thus, the arm was in a closed container. Any change in volume of the arm, as by an increase of blood flow, would push the rubber cover out slightly, and this would be reflected by a deflection via the lever onto the smoked drum. The technical term is a plethysmograph, and it is the basis of a lie detector. The volunteer sat on a stool with his encased arm on the lab bench. The examiner attempted to stress the student by asking mental arithmetic questions, but even when expressing mock surprise at a correct answer, he was getting no response from the placid Nigerian guinea pig. In desperation, he produced from behind his back a starting pistol, which he suddenly discharged. The student leapt up and shot off, taking the apparatus, smoked drum and all, with him.

Another student sat calm and unperturbed, the smoked drum showing a horizontal line. Eventually, one of his so-called friends shied up to the lecturer and whispered something in his ear. With a smirk of triumph, the lecturer then spoke.

'Oh, by the way, what were you doing on Sunday afternoon on St Mary's island?'—and the lever shot up vertically within seconds!

As you will have guessed, he had been indulging in a romantic liaison.

Chapter Six

The Bun Room

About halfway through the Easter term, Eric Barkworth—who I had recognised on day one as having been suspended from school for wearing a duffle coat—told me about a vacancy at his lodgings with Mr and Mrs Stuart in Brighton Grove, in the west end of Newcastle. Mrs Neylon was very sweet about me leaving, saying she had noticed I was unsettled. The Stuarts were a jolly couple. George Stuart had been a petty officer in the Navy and sometimes bounced into our room when we were still in bed, calling us to rise and shine and performing energetic callisthenics, even though he was now middle-aged, overweight and a smoker.

Eric was a great guy, permanently cheerful and bright-eyed, with fine angular features and slim, with jet black Brylcreemed hair and a slight bounce to his walk. I met his girlfriend once when she greeted him at Paragon Station, jumping up and wrapping her legs around his waist and she was a beauty, appropriately named Elizabeth Taylor, like the film star. His affection for her did not stop his roving eye, however. Once we were having lunch in a Chinese restaurant when he spied a girl a few tables away and started giving her the eye. Finally, he wandered across, smiled, asked her out and got her phone number. He returned as cool as ever, as I sat there open-mouthed. He had a completely different background to me, coming from a wealthy part of Hull I scarcely knew existed. He openly volunteered that he was studying medicine because of pressure from his mother.

One evening when it was dark, Mr Stuart came to our room and whispered for us to follow him to the upstairs landing, where there was a window looking down on his neighbour's kitchen. The neighbour, a dark-haired lady in her forties, stood in front of the fireplace and stripped completely naked, and then appeared to admire herself in a mirror beyond our vision, before leaving the room. Mr Stuart had obviously witnessed the performance before, because it was almost as if she did it on cue. What shocked me at the time was not only that he called on us to witness the act, but that it was with his wife's connivance.

Mrs Stuart occasionally took in 'theatricals', and when the Royal Ballet appeared at the Theatre Royal, one of the cast, Elizabeth Anderton, stayed, as well as a male choreographer. Being mature for our age we placed a skull containing a lighted candle in the toilet to frighten her, but she simply laughed. The only time I saw Miss Anderton was from behind when she walked along the landing, and she seemed to float. She later was a *répétiteur* with the Royal Festival Ballet and when an Achilles tendon injury ended her career as a dancer, she taught with Sadler's Wells.

We were still in our first year and didn't yet feel part of the culture of the North-East. Although we occasionally visited city-centre pubs, we tended to confine our drinking to the Bun Room in the basement of the Students' Union building. As well as toilets, the cloakroom had baths, and for ninepence in old money one could have a hot bath plus the use of a long brush for scrubbing one's back, and a whole bar of green soap. This is the only venue in which I can recall taking a bath until I qualified. When we later moved into rented flats the baths were unusable, being bereft of enamel.

The advantage of the Bun Room was that one could meet fellow students and sing raucous, filthy songs, and possibly go to a dance afterwards. The songs were standard throughout every university campus and rugby club in the country—*The Engineer's Song, The Ball of Killimuir, The Wild West Show* and a version of *The Twelve Days of Christmas* in which every verse ended with 'and my Lord Montague of Beaulieu'. Closing time was ten o'clock, later extended to ten-thirty, with ten minutes drinking-up time, when there would be a conga line leaving the premises to the strains of, 'Lloyd George knew my father, father knew Lloyd George'. All very childish. It wouldn't be allowed today.

The dances reflected the male-dominated society of the time and I feel embarrassed thinking about them. The usual routine was to sink a few pints in the Bun Room and at closing time, go next door to the dance. The music was traditional ballroom, with the occasional jive, and later on, the twist. Even shy, diffident souls such as myself, probably emboldened by alcohol, would slowly parade up and down, assessing the remaining 'talent'. I once asked a pretty girl in a floral dress to dance. At the end of one number she lingered for the music to commence for the next dance—an encouraging sign, I thought. Halfway through the dance, a familiar Somerset voice called out, 'Aye Dive, 'ow yer doin'?' Before I could answer, she asked if I knew the speaker.

'Yes,' I said. 'He's a friend of mine.'

In a flash, she was gone, and I have never seen her since. I was guilty by association. The culprit was one Nicholas Alcwyn Wright of the year above, but actually younger than me because he won a state scholarship at the age of sixteen. Nick was one of those well-disciplined characters who worked hard during the week and played hard at the weekend. He was of slightly below average height but broad and strong, slightly bandy-legged with wiry wavy hair, a broad Somerset accent and a way with the ladies.

During that first year, a few of the lads grew beards. Professor Scothorne remarked that this was a perfectly natural thing to do, to prove one's manhood, but that after 2nd MB, when hospital practice commenced, they should be shaved off. Quite right too!

This was about the time of the publication of the Royal College of Physicians' report into the relationship between smoking and cancer of the lung.

'As from today, you will cease the smoking of cigarettes,' Scothorne told us.

And then with a little smile and a glint in his eye, he added, 'I, of course, shall continue to smoke cigars.'

Cigars were excluded from the report—presumably because in general, cigar smoke was not inhaled, or possibly there were too few cigar smokers to be of statistical significance.

***

In the Medical School, the sexes had separate common rooms. With a five-year course there was a wider age range, made even more so by the recent abolition of National Service. Some final-year students were more like men compared to us, with moustaches, some already balding, wearing waistcoats and watch chains, smoking pipes as opposed to cigarettes and playing serious card games. Perhaps apocryphally, one chap was said to have either won or lost, I can't remember which, a Jaguar car. His eyesight was so poor that I doubt he was able to drive. His skill at cards was put down to the proximity of his hand to the lenses of his spectacles, which looked like the windows of a Victorian sweet shop. Colleagues joked that the only member of the family without glasses was the dog. He inherited the condition from his father, a GP, who must have been able to drive. He once approached me in the library and asked if I was one of his friends. At first I thought he meant was I friend or foe, but soon realised that he could not make out my features. Rumour was he once sutured his mask to a woman's perineum, which he was repairing after childbirth.

Friday night meetings of the Medical Society gave us the only opportunity in the pre-clinical years to venture across the road into the hospital. The main entrance had a beautiful mahogany-panelled vestibule, with the head porter's lodge and to one side the consultants' dining room, and on the other a huge grandfather clock at the foot of winding stairs leading to the boardroom with its long table and oil paintings of past dignitaries. Finally came the doctors' dining room for junior staff. Walking past this atrium, the main corridor was two to three hundred yards long and about halfway down on the left was the New Lecture Theatre where the meetings were held.

A few meetings during our first year stand out. One was an address by Sir Zachary Cope, a wonderful name for a surgeon. He wrote a short classic, *The Early Diagnosis of the Acute Abdomen*, first published in 1921. He was a small man of about eighty and the only thing I remember was a fellow student, Sheila Saville, remarking as we dispersed afterwards, 'He didn't seem to have the hands of a surgeon.' Rather like many of the public, we thought there should be something different about the configuration of a surgeon's hands. Then there was Albert Sabin, an elderly, white-haired, cheerful American, who developed the oral polio vaccine, and was world famous, the medical equivalent of viewing a film star.

Another was the nameless psychiatrist who seemed to specialise in treating mental illness in clergymen. He described interviews with patients under the influence of lysergic acid (LSD) in which they recalled in detail past-life events, including a man whose apparent first experience was a tremendous headache, and it transpired that he had been born into a toilet bowl. LSD was unknown to us, becoming a familiar term only years later.

We also had a demonstration of what was termed 'natural childbirth' by an Irish doctor who was a disciple of Dr Grantly Dick-Read, who sold millions of copies of *Childbirth without Fear: The Principles and Practice of Natural Child-birth*, but was viewed with suspicion by many in the profession and is almost forgotten today. He hypnotised a lady who was expecting a child shortly and used this medium to relieve the pain of childbirth. This was the first time I had seen the technique live and it was most impressive. He made several suggestions to his subject, such as that her feet were on fire, and she responded accordingly. Feedback at a later date was disappointing because she needed a forceps delivery under general anaesthesia.

Some time during the summer term of 1962, Eric shared doubts about continuing with medicine, although he didn't have an alternative in mind. He went to see his tutor, Professor Scothorne, whose unusual advice was to read the chapter on the nervous system in Wilfrid Le Gros Clark's textbook *The Tissues of the Body* and then return to see him. Eric obediently waded through this seventy-plus-paged chapter, occasionally smiling and commenting on how interesting it was, but to what end? When he expressed this view to the prof on his second visit, he was told that if he hadn't found it truly inspiring then he probably wasn't cut out for medicine, but to ponder further on it. One tends to think mature intellectuals have the answer to all such problems, but clearly the prof was leaving it to Eric, albeit with a hint. At the time I thought the advice odd, but the book was a classic and the author had the ability to make what superficially appeared a dull subject absolutely fascinating, and I now think it was a novel suggestion. Eric did give up the course and I never saw him again. I expected him to become an entrepreneur, but one of our classmates, a credible witness, claims to have seen him working as a manager at Mothercare in Glasgow.

So in Michaelmas term, 1962, there was all-change at Brighton Grove. I was joined by Roger Gomersall, whose landlady in the adjacent street was relinquishing the business and knew Mrs Stuart. Within a short time Forbes was asked to leave his digs, so Mrs S took pity on him and the three of us slept in the large front bedroom, while a new lad, Barry, from Worthing, had his own room next door. He was in his first year, studying physics. He was blond and red-cheeked and already wore very thick lenses in his glasses. We never suspected a thing but within only a few weeks Barry was diagnosed as diabetic. One morning he was unwell, without our knowledge, as we had already left for lectures. Mrs Stuart contacted her GP, a family friend, who immediately diagnosed the problem. He wasn't admitted to hospital but treated with insulin and initially managed as a daily outpatient until stable. In his absence, Mrs Stuart showed us a cupboard in his room stacked with empty barley water bottles. Apparently, he asked for a jug of iced water every night from her, so he must have had the terrible thirst typical of the condition. All this passed us by—not that we knew anything about diabetes anyway.

One afternoon in October 1962 I entered the Barras Bridge refectory for lunch and joined my classmate Keith Whaley, who looked quite worried.

'Do you realise that by four o'clock this afternoon we could be at war?' he said.

I thought he'd gone a bit cracked until he related the business about the Russian missiles heading towards Cuba. Even then, being young and daft and with little knowledge or interest in politics, I was loath to take him seriously. But of course, it was true. In my circle, none of us read a newspaper and we had no access to television.

Keith was a bright individual who became a Professor of Immunology, first at Glasgow and later Leicester, but his appearance belied his intellect. He was rugged, with prominent supraorbital ridges, a flat face with uneven teeth, and he had hands like spades. Naturally, he played rugby. The event I am about to describe occurred in 1965, when he was playing for our team, fifth year, against final year in the traditional annual match, usually a torrid affair. During a maul which was getting nowhere and with the whereabouts of the ball unknown, the ref blew his whistle. As the combatants separated and assumed an upright position, Keith was last to rise from the bottom of the pile. A member of the final year, the son of a prominent paediatrician, aimed a kick at Keith's head as he retreated. There was a sickening thud as he connected and the onlookers sucked in their breath in shock and anticipation. Keith merely shook his head and looked up. After a few seconds, some joker broke the silence.

'Just as well you kicked him in the head,' he said. 'Otherwise you might have killed him.'

Forbes, who had now joined me in my digs, continued to frequent the same pub at lunchtime and would yarn about the characters he met. He became infatuated with a married woman called Leah and was quite open in conversation around the table at our evening meal about his love for her, so much so that Mrs Stuart got wind of it. When it was briefly mentioned that Leah's husband owned several tobacconists' kiosks in Newcastle, the penny dropped and she told Roger and me to warn Forbes to terminate any relationship. Apparently, the husband, who was considerably older than Leah, was a black marketeer during the war and with his criminal connections, was likely to have Forbes harmed. We did wonder what charms the mysterious Leah possessed. One afternoon we found out. We were in the dissecting room when Forbes suddenly appeared late. Two men and a woman, all wearing white coats, accompanied him. They did not appear to be drunk, but had certainly been imbibing, and alcohol must have played some role in emboldening them to enter such a place, despite Forbes' obvious invitation. Instead of shooing them out, the demonstrators immediately left the room, the cowards. One introduced the other as a kidney transplant surgeon from Glasgow, which seemed implausible for two reasons. Firstly, he was scruffy, unshaven and had grubby, bitten-down fingernails. Secondly, kidney transplants had not yet taken off in 1962, except for identical twins, because immunosuppressive drugs were only introduced that year. The trouble with Forbes was that he was often as gullible as he expected others to be—although to be fair, some of his more outlandish stories proved to be true. He certainly led a colourful life. The woman, naturally, was Leah. She had black hair, flattened against her scalp rather like a 1920s style, scarred cheeks from bad adolescent acne, blue eye-shadow and lipstick applied so badly that it appeared contrived. She looked like I imagined a low-class lady of the night would appear. To give her credit she was unfazed by the company of several cadavers, nor was our transplant surgeon. In reality he was a lorry driver, but he gamely attempted to explain where certain anatomical structures were situated when asked, apparently quite seriously, by a, 'Sir, sir, can you show me where the femoral nerve is?' Grasping a hook from the questioner, he poked it about before elevating a piece of tissue and declaring that it was 'in there somewhere'.

The fiasco was soon over, thankfully, because although it was entertaining, it was quite wrong. Forbes could well have found himself out on his ear if a demonstrator had not returned and told them to skedaddle as the prof was in the vicinity.

The winter of 1962/3 was one of the worst on record. The temperature rose little above freezing until the end of March, and snow was piled up a few feet high by the roadsides for weeks. It was around this time that we suddenly found a stranger in our midst, Niall Cartlidge, who has remained a good friend. Niall was the proud owner of a minivan and had been thrown out of the University of St Andrews for attempting to be the getaway driver for a couple of fellow undergraduates who intended to let off the fire alarm in a female hall of residence. Unfortunately for them, the university police interrupted and reported them to a higher authority. Despite failing in their mission, they were dismissed, and even an appeal to the rector, the author CP Snow. A long journey to visit him in London, had no effect. The rector and his wife, the poet Pamela Hansford Johnson, were sympathetic and fed them tea and biscuits, but Snow admitted he was powerless.[1]

One of Niall's tutors was impressed by his ability and wrote to Professor AGR Lowdon, dean of medicine at Newcastle. Niall tells of sitting outside the dean's office, trembling with apprehension, when Professor Lowdon approached. A little short of breath, he admitted him to his study, apologised that he was late for an appointment and simply said, 'You're in' and shook his hand. It was St Andrews' loss and our gain. Not only was he the first person to achieve first-class honours in medicine with a distinction in medicine and surgery for two decades, but he became a Professor of Neurology in the university. Life and fate!

---

1. Snow was well-known, mainly for his essay On Two Cultures, but was not well regarded as a novelist. He studied natural sciences at Oxford and initially worked as a physicist before turning to writing novels. In his discussions with the literary set he was struck by the fact that while they regarded scientists as philistines, they had hardly any knowledge of science themselves. This gulf in knowledge between the sciences and the arts seems to derive from the historical stress in England on the classics in education. With the obvious striking exception of Newton, few of the pioneers of science in this country had a university education, and, of course, until the 1830s there were only two such institutions.

Chapter Seven

The Need for Speed

S pring 1963 was the time of the dreaded 2nd MB hurdle. Providing you passed, you were unlikely to be subsequently thrown out, so a period of intensive revising took place as it approached. One of the main causes for concern was the three-hour anatomy written paper. Questions might take the form of, 'Describe the course, relations and supply of the median nerve in the forearm', or 'of the femoral artery in the thigh'. Such anatomy is easier to describe *in situ* in an oral exam than imagined from memory, which is a totally artificial situation. After all, one only needs the knowledge when the anatomy is visually exposed, either by trauma or electively during surgery. Nevertheless, one had to play by the rules. Between the three of us we calculated that if we worked through the night we could just about complete a quick revision, sit the exam and return home to sleep. Mrs Stuart was in on the plan. She provided a couple of flasks of black coffee and her masterstroke, amphetamine pills, which she had been prescribed for weight loss. Privately, she told me and Roger she only had two tablets left but would give Forbes an alternative and to keep schtum. We worked all day except for meals, retired at ten, and set the alarm for two in the morning. Young people can go without sleep easily, but on being roused we did wonder if this was a good idea. Anyway, we took our pills, making sure Forbes didn't see them, and poured out the coffee. A few minutes later my head felt as if it had been thrust back and everything seemed hunky-dory. Obviously, the uppers had kicked in, and we carried on as planned. There was never any comment from Forbes, to my recollection. Later, we discovered he had been given a De Witt's kidney pill, which contains the dye methylene blue and turns your pee green!

In connection with this last-minute swot our demonstrators introduced us to mnemonics, which had probably been passed down since time immemorial. The cleanest one was that for the twelve cranial nerves—'On Old Olympus' Towering Tops, A Finn And German Picked Some Hops.' That stood for olfactory, optic, oculomotor, trochlear, trigeminal, abducens, facial, auditory, glossopharyngeal, pneumogastric (the old term for the vagus), spinal accessory and hypoglossal. The branches of the external carotid artery in the neck—ascending pharyngeal, superior thyroid, lingual, facial, occipital, posterior auricular, superficial temporal and internal maxillary, were represented by, 'As she lay flat, Oscar's penis slipped in.' The problem was, of course, remembering what the particular mnemonic stood for. For example, my favourite was not a mnemonic but a little rhyme—'The lingual nerve took a swerve, around the hyoglossus; well, I'll be f****d, said Wharton's duct, the bugger's double-crossed us.' I cannot for the life of me recall what, 'Lazy French tarts sit naked in anticipation' was supposed to help with.

Well, I satisfied the examiners, to use the formal expression, and the realisation now came that, barring disasters, I could be a medical practitioner in about three and a half years' time. Those who passed 2nd MB now started an introductory clinical course in the hospital, while the failures resat after a break and a period of revision. Subsequent failures had to leave.

After the summer, George, Jim, Roger and I planned to get our own flat. From now onwards, unlike other undergraduates, we would attend for about forty-six weeks of the year. This meant that we would have to leave the Stuarts, a sadness because we almost felt part of the family. Mrs Stuart looked after us like a mother—in fact, I once inadvertently called her 'Mother' and although I immediately apologised, she glowed with pride. However, she needed a proper break during the summer, so the separation was inevitable and, of course, we would have more freedom, although poorer food and a colder environment.

In 1984 the Medical School celebrated its one hundred and fiftieth anniversary with a two-day *Festschrift*. I attended and met Dave, the dashing best friend of the

Stuarts' son-in-law, Bill, and who was now an obstetrician in Canada. During our reminiscences he told me Mr Stuart had died on a massage table at the Turkish baths he was always encouraging us to attend. He then astonished me with the revelation that Mr Stuart used the venue for years as a meeting place while he acted as a fence, which probably explained his knowledge of Leah's husband. Sadly, Dave the handsome hulk died suddenly of a heart attack after skiing, at only sixty years of age.

For my friend George Bone and myself, our temporary accommodation over the summer of '63 was in the Highbury area, on the top floor of a large house that belonged to a Dr McEwen and his wife, an elegant, grey-haired lady who had obviously been a beauty in her youth. We didn't have much conversation with the doctor, who was a geriatrician in Sunderland. He had clearly suffered a stroke, as he had a paralysed arm and swung the leg on the same side from the hip. As a young man, Dr McEwen was operated on by George Grey-Turner for a carotid body tumour, a very rare condition. Such tumours are almost always benign, but do grow progressively and can create local symptoms. They have a very rich blood supply and their plane of separation from the parent carotid artery can be difficult to identify. It seems that even such a surgical giant (figuratively speaking; he was in fact a small man) as Grey-Turner had difficulty because he had to resort to ligating the internal carotid artery, which with its contralateral companion, forms two of the four arteries supplying the brain. In about a third of cases following such ligation a permanent stroke results, because the cross-circulation from the other side is inadequate. Grey-Turner was a Newcastle surgical legend in his own lifetime, but he left the city in 1934 to become the first Professor of Surgery at the new Postgraduate Medical School at the Hammersmith Hospital in London, so clearly Dr McEwen's operation was earlier than this.

Before George and I made our first tentative steps across the Queen Victoria Road to the RVI, we had to purchase our badge of office, a Sprague-Bowles stethoscope, placed to poke out of the pocket of our crisp white coats. We never, but never, wore it round the neck—that was for senior staff only. The modern conceit of wearing the instrument behind the neck and across the shoulders owes itself to the lighter Littmann model and a plethora of hospital TV soaps.

We were assigned to different 'firms'; me to Dr Boon's on wards ten and twelve, and George to Dr Dewar's on wards fourteen and fifteen. Since the European Working Time Directive was extended to cover junior doctors in 2004, the term 'firm' has been obsolete. It usually referred to a team of two consultants, a senior registrar, a registrar, sometimes a senior house officer and two pre-registration housemen, who together were in overall charge of two wards, male and female, of about thirty-two beds each. The pre-registration housemen (the females were also referred to as such) were newly qualified and by law were required to do at least a year at the grade, in medicine and surgery, for six months each. After this, one's name was placed on the medical register and the world was your oyster, so to speak.

Senior house officers had just completed their house jobs and were in post for a year or more. If they wished to enter general practice, many would do jobs in paediatrics and obstetrics first. Others would flit around until they found the specialty they wished to pursue. The registrar was a training post in that particular specialty leading to the relevant postgraduate diploma, and occupants were in post for three or more years. The senior registrar was the final rung of the training ladder before applying for a consultant post. Depending on the specialty and availability of vacancies, this might be held for between three and six years, and exceptionally even more.

Virtually all the teaching on my introductory course on wards ten and twelve was done by the SR, Chris Good, and the senior house officer, one Miriam Stern. Chris was in appearance a slimmer version of one of my boyhood heroes, Peter May, the England cricket captain in the late 1950s. Miriam later found national fame as an author, journalist and TV personality on matters of health, under the surname Stoppard, after her marriage to the renowned playwright and is now Lady Hogg. Miriam was obviously bright and keen to teach. She had a pretty face, was slightly plump, and she knew how to make the best of herself. She was always well dressed and tastefully made up. Her black hair reminded me of of

Elizabeth Taylor's *Cleopatra*. She bustled around, delighting in taking a seminar perched on a high stool, swinging her legs from side to side, her skirt above her knees and somehow mesmerising us to focus on them. Then when we looked up she regarded us with an amused, flirtatious smile. Or am I imagining this? No, I'm not!

Her enthusiasm and possible naivety got the better of her, and us, when one afternoon she introduced us to a 'paediatric biochemist' whom she had just met and invited to attend an out-patient clinic. I have forgotten his name (or at least the one he used), but he had an impressive European-American accent, the sort that develops in emigrants to the USA from eastern Europe. Miriam aimed all her questions at us students, courteously letting the newcomer off the hook, but she asked him to listen to heart murmurs, which he appeared to do satisfactorily. He appeared regularly in the students' café in his white coat and stethoscope and seemed to us tyros to be entirely plausible, although to my knowledge no-one questioned him about his background or credentials. Even the label 'paediatric biochemist' was listed in the medical school prospectus.

However, one afternoon I was outside the porter's lodge scanning the medical school noticeboard when I spotted a letter for him. Commenting to a companion as to whether I should take it for him, the hand of Bert, the porter, was placed on my shoulder. He swiftly ushered me into the office of the registrar, Norman Shott, who asked what I knew about this chap. After my short account, he leaned back in his chair and drawled in his RAF manner.

'That's absolute balls,' he said. 'He's obviously a con man.'

But he thanked me and I left. That evening, a certain Jimmy something or other, who had never travelled much further than Heaton, in Newcastle, where he lived with his now distraught widowed mother, was arrested. We had never associated his presence over a few weeks with a series of petty thefts of cash and valuables from cloakrooms, perhaps because they advised the victims to keep quiet until they apprehended the culprit. Unfortunately, one student foolishly signed his name on the front of his chequebook, which was stolen. I believe our paediatric biochemist got two years. What a waste of his undoubted talent.

Although I never knew her in any capacity other than student-teacher and hence cannot confirm it, only she could have been the original source of another story involving Miriam. Stern was her maiden name and before she married Tom Stoppard she had an earlier marriage. Apparently, she and her first husband were holding a party one evening. One guest was a weird chap in the year above us, who attended with his long-term girlfriend, Judy. He was an unlikely sex god, being bespectacled with an unruly mop of curly blond hair, a more or less permanent vacuous grin and a sploater-footed gait. Anyway, he and Judy turn up for Miriam's party much earlier than expected. Miriam answers the door, is surprised, but tactfully explains that she and her husband are still preparing the food, and if she can be excused, they can wait in the front room with a drink and she will join them in due course. Only a few minutes later Miriam pops back to be sociable and see if another drink is required, only to find the couple *in flagrante delicto* on the lambswool rug in front of the fire!

***

We were taught how to take a structured history from patients and conduct a physical examination. The standard textbook was *Hutchison's Clinical Methods*. Initially, six of us would gather round a patient in bed, curtained off from the rest, and our teacher would demonstrate the various steps in the procedure. Gradually, we were let loose, in pairs, one taking the history, the other doing the examination. We were more nervous than the patients, who often seemed to take some pride in being chosen, or having some relief from boredom, or being of some educational value. I can honestly say that in all the years I spent in the RVI, both as student and trainee surgeon, I cannot recall a patient ever refusing to be a 'guinea pig', one of the many reasons I am so fond of the Geordies.

The wards in the RVI were of the Nightingale design. The entrance to each had a set of double doors which were only closed at night. They led first to offices for consultants and secretaries, a seminar room, treatment or dressings room, toilets, storage and, just before the ward itself, Sister's office, which usually had a fireplace and in winter, a coal fire. The parquet-floored ward space itself started with four cubicles, two on each side, followed by twelve beds on either side, and at the very end two more cubicles, one on each side, and beyond that the patients' toilets and the sluice, where bedpans and metal male urinals were washed and sterilised in an autoclave. Student nurses wore a blue and white pinstriped dress, starched plain hat with a horizontal blue stripe and white aprons. Staff nurses had plain grey dresses, but the sisters wore a lovely shade of mauve.

As students, we were responsible for side-room testing of urine and faeces on a rota system. In 1963, as now, urine was routinely tested for glucose, protein, bile and microscopic blood using proprietary paper sticks. Suspicious-looking urine was also subjected to microscopic examination for pus cells (indicating infection), crystals and 'casts', which might indicate kidney disease. Stools were tested for occult (hidden) blood, that is, blood invisible to the naked eye, and which might indicate a bleed somewhere in the gastrointestinal tract. A significant amount would turn the stool tarry black, so-called melaena, but iron tablets could give a similar appearance. Testing was done using a smear of faeces on blotting paper and adding reagents from bottles, hydrogen peroxide and ortho-toluidine, which turned blue in the presence of blood. This is the principle behind the national Bowel Cancer Screening Programme.

Dr Tom Boon was a general physician with an interest in gastroenterology, although that sub-specialty hardly featured in the days before the invention of the fibre optic instrument by the unsung hero Harold Hopkins. The only way Tom Boon could directly visualise the stomach was by a rigid brass gastroscope, which had to be passed under a general anaesthetic unless the patient was a sword swallower, and it had several blind spots to boot. But his interest provided us with another duty, that of gastric cytology. Patients with pernicious anaemia cannot produce acid in the stomach, the lining of which has a form of gastritis which predisposes to an increased risk of cancer about three times the normal rate. They advised such patients to attend every six months and have a plastic tube inserted via a nostril into the stomach. Some juice was then aspirated and sent to the pathology lab to be centrifuged and examined microscopically for cancerous cells. Understandably, they dislike the procedure, and there is an art to passing a nasogastric tube. It was a task we did not welcome either, especially as sometimes up to six patients would turn up.

I can only remember Dr Boon teaching us once. He approached Chris Good one day and said he was free and would teach us, and an appreciative SR shot off before he could change his mind. He took us to a patient with an enlarged spleen and made us all get chairs and sit down so that the bed curtains spread out over our backs. He then borrowed a copy of *Clinical Methods* and read out the section on the spleen, before showing us how to feel for the organ, telling us each to take our turn. After that he bade us good day. He examined me for my insurance policy a few years later when we needed a mortgage, seeing me in his consulting room at home.

'You haven't had VD yet, have you?' he said, as he went through a checklist.

It was the 'yet' that amused me.

***

We now had a relatively relaxing time, with a combination of lectures and hospital practice, and no examinations until spring 1964, in Pharmacology and Materia Medica, to give its full title. This subject was taught solely by Tophie Wynn, a short, tubby man in a baggy suit, with a walrus moustache and apparently sad demeanour, but who was, in reality, a pleasant man with a droll sense of humour. Of middle age, he was the last person to achieve first-class honours in medicine before my friend Niall Cartlidge. He lectured in a slow, flat delivery, apparently without any interest, but he knew his stuff. The message from the year above was to not miss his last lecture before the exam because he would give advice on how to set out an answer to a question.

'Let's say, for example, you were asked to discuss iron compounds in therapeutics,' he announced in his seemingly bored monotone. 'How would you start?'

He then spelt out on the blackboard exactly how to do so and, sure enough, when we turned over the examination paper a week or so hence, there it was.

We also started lectures and demonstrations in pathology, bacteriology and public health, but the examination in these subjects was not until the summer of 1965. Pathology, or the study of disease processes, is the basis of all medicine and together with surgery, it was my favourite subject.

We started with general pathology, which covered acute and chronic inflammation (the latter exemplified by tuberculosis), thrombosis and arteriosclerosis and malignancy. Lectures were complemented by microscopic slide demonstrations on an old-fashioned sparking projector, the mode of function of which I could never decipher—very different from PowerPoint. This was followed by personal study of our own numbered slides in their varnished wooden box.

The main lecturer was the Professor himself, Alfred Gordon Heppleston, a tall, upright, very precise and humourless Mancunian. His own field of research was the pathogenesis (that is, the process creating the disease) of emphysema, and he was certainly in the right place for that, the Northumbrian and Durham pits being still active in the 1960s. Heppleston was immensely proud of the extensive museum he had created, with preserved organs in Perspex cases lining one entire side of the laboratory wall.

To fulfil the requirements of the course, we had to attend a minimum of twenty post-mortems. Naturally, the post-mortem room was adjacent to the mortuary, with its metal containers for the bodies, which slid in and out of a giant refrigerator. With the strange black humour present throughout hospitals, these containers were, and still are, referred to as 'chip pans'. There were tables for three autopsies to be performed simultaneously, separated from the spectators by a solid wall at about chest height, with rows of tiered steps for observation. Just over the wall was the marble surface on which the pathologist demonstrated the abnormalities in the various organs, because we were not permitted to go beyond it. Perched on the wall were several metal ashtrays. Not only was smoking allowed, but even the non-smokers also encouraged it, to partially offset the sickly-sweet smell created when the bodies were opened. I suspect they were there also for the benefit of the police who attended coroner's cases, and most of whom smoked. When the ashtrays were full they were simply emptied into the abdomen of a body before closure. This distasteful act shocked we students as being somewhat disrespectful.

Coroner's cases were conducted by the Home Office pathologist Colin Corby. Despite the association of this title with murders and detective fiction, and in modern times, TV dramas, the vast majority of these deaths were due to natural causes. The other type, so-called hospital post-mortems, were usually performed by junior pathologists. At my first attendance I was surprised to see a mousta-chioed man with naval tattoos on both forearms opening the body, and thought it strange that a medical practitioner would display such emblems. Eventually, it dawned on me that he was a mortician, whose job it was to prepare the body for the pathologist and sew it up afterwards. There were two of these men, whom we got to know. One Monday the aforementioned mortician came in to see what workload had been created over the weekend and opened one casket to reveal the face of his colleague, Walter.

The morticians were especially skilful in removing the brain. An incision was made behind the ears and the scalp was reflected forward over the face. The vault of the skull was elevated with a circular saw and the brain removed for exami-nation. It was then replaced and the scalp stitched back so that no disturbance would be visible if a relative wished to view the deceased before the funeral. Autopsies have been common throughout history and the results have often been made public, even in the cases of royalty. In the 1960s such 'interest' autopsies outnumbered coroners' cases. As a junior doctor, we were frequently delegated with the task of requesting one from a bereaved next of kin. It was an unpleasant job, but it was rarely refused, the comment usually being along the lines of, 'If it will help you, doctor, or if it will help someone else...'

Incidentally, with reference above to the brain, morticians usually removed the pituitary gland, not much bigger than a pea, which sits at the base of the brain in its own little *fossa*. Rumour was that they received a small sum for collecting them on behalf of a pharmaceutical company. In those days, before the advent of molecular biology and genetic engineering, the only source of growth hormone to treat the rare condition of congenital dwarfism was from cadaveric pituitary glands. Growth hormone is one of several hormones secreted by the pituitary gland, which has been labelled 'the conductor of the endocrine orchestra'. When human CJD (or mad cow disease) hit the headlines a few years ago, some cases were attributed to this practice, in which a large number of glands were pooled to produce enough growth hormone for treatment. Presumably, at least one must have been the source of the causative agent, a strange entity called a protein prion, which is not destroyed by standard methods of sterilisation.

'Interest' post-mortems are almost unheard of today. There are probably two main reasons. Firstly, with modern sophisticated diagnostic aids, such as CT and MR scanning, it's unusual to find an incorrect pre-mortem diagnosis, excluding cases of sudden death, which are referred to the coroner. Secondly, adverse publicity was created by the so-called 'stripping of organs' at Alder Hey. Unfortunately, the whole business was badly handled by people who should have known better.

Pathology included clinical chemistry, the departmental head being Professor Albert Latner, an ebullient Cockney Jew with wavy black hair, Roman nose and the retained accent. He was once lecturing on diabetes and discussing the treatment of hypoglycaemia (low blood glucose) in the situation of a GP, using what he might have in his bag.

'Some people say that an injection of adrenaline will raise the blood sugar,' he said. 'To use my erudite and professorial expression, this is cock! The adrenals are already pouring out sufficient adrenaline—yours will make no difference.'

In a *viva*, he asked a lad a possible cause of melaena (faeces containing partly digested blood).

'DU,' the candidate replied, meaning a duodenal ulcer.

'DU? DU? Don't you DU me, my lad—for all I know, DU could mean Dirty Underpants!'

The lesson was, do not use abbreviations in an examination situation.

It's interesting how bacteriology has changed. Even the terminology of some bacteria has altered due to changes in classification. The large intestine is full of anaerobic bacteria (Bacteroides)—anaerobic referring to their ability to survive in the absence of oxygen. In the sixties they were thought to be harmless companions of the better-known *E Coli*, but for some time they have been recognised as pathogenic under certain conditions. *Shigella sonnei* was a common cause of diarrhoea but is virtually unknown now, whereas *Campylobacter jejuni*, of which we had no knowledge but which inhabits the guts of birds, is now a common cause of the illness. This is possibly because chicken is now the cheapest source of meat protein and is often inadequately cooked, especially in the summer, when the barbecue seems to be the preserve of the male of the species, who tends to be less conscious of kitchen hygiene than the mother of his children.

Much attention was concentrated on the properties and culture media requirements of the bacteria, which were of less importance to us than the diseases for which they were responsible. We received tuition in public health from the Medical Officer of Health for Newcastle, Dr Pearson. I looked forward to this, thinking we would learn about epidemics and their control and issues such as sewage disposal. But it's fair to say that none of us was on the same wavelength as our esteemed lecturer. There was no structure to the course and it was pointless to take notes. He seemed to ramble from one thing to another haphazardly, although on reflection I think he was a very intelligent man trying, but failing, to persuade us to think laterally. One of the few pearls that I recollect is that in designing a lift for a tower block, one must ensure that it will accommodate a coffin lying flat.

Chapter Eight

Firm Friends

After that pleasant summer of 1963 on the introductory course, George and I had to move out of Dr McEwan's house in Highbury. In October, at the beginning of Michaelmas Term, we moved with Roger and Jim Turley into a flat in Myrtle Grove in Jesmond. Our landlord was a Mr Beecham, who lived nearby in more upmarket Towers Avenue. The Beechams were Jewish and their son, Jeremy, sat at the table sipping soup while wearing his kippa cap, as his mother exercised her platysma muscle to a degree that I had never witnessed before nor since. This is a subcutaneous muscle in the neck, the human equivalent of a horse's withers, which becomes pronounced when one stretches the mouth in a rictus smile with downturned lips. It must have been a nervous tic, for it occurred every few seconds. Mr Beecham insisted our flat was 'a palace' compared to what his son had in Oxford. If that was the case then Jeremy warranted our sympathy. He must have made his parents very proud because he became a lawyer and ultimately the Labour leader of Newcastle City Council as Sir Jeremy and finally, Lord Beecham.

The houses in Myrtle Grove were terraced, with backyards and front doors that opened onto the street. Backstairs took us down to the toilet in the yard. There was a bathroom but the bath no longer possessed its white enamel, the brown undersurface being exposed, and hence it was never used. The flat was spacious enough. We each had a bedroom and there was a living room and a small kitchen. The décor was a depressing dark brown throughout.

Lunch was always a snack in the hospital café or university refectory. Evening meals were haphazard. Sometimes on a weekend we would all muck in and produce a stew and mash. We had a short enthusiasm for cheap meats such as heart and shin beef, and Carthorse (as Niall became known) would plunge a cow's femur into the large black pan to add to the flavour. I'm not sure how many meals this osseous ingredient assisted, but my recollection is that it was reused for weeks. During our medical and surgical 'clerkships', each firm was in a rota for emergency admissions, roughly one day in four. Prospective admissions were phoned in by the GP to either the RMO (Resident Medical Officer) in the case of 'medical' cases, such as heart attacks, asthmatic crises and chest infections, or the RSO (Resident Surgical Officer) for surgical cases, usually encompassed by the term 'acute abdomen', such as appendicitis, intestinal obstruction and perforated ulcers. Between the hours of nine to five, these individuals were housed in their respective 'cabins' in the Accident Room (never referred to as 'Casualty' in the RVI), where they acted as first assessor of such patients, requesting baseline investigations and x-rays and deploying them to the ward on reception if admission was deemed necessary. These officers were of registrar grade and their role in semi-independent practice gave them valuable experience, as I later learned. Of course, they could always obtain the opinion of the senior registrar, or sometimes the consultant, before formal admission to a ward. Once admitted, the permanent team took over. After 5pm, all calls via the switchboard then went to the registrar or SR of the firm in question.[1]

Before final year we were attached to medical and surgical firms in the mornings and following afternoon lectures we returned to the wards for the 'reception ward round', conducted by the consultant and his team. All the emergencies were presented to him by the houseman and the diagnosis and plan of action were debated. We students, of whom there were up to six, would either observe the proceedings or one would be chosen to take a history and another to undertake the examination, supervised by the consultant. The round started at about 5.30 pm and usually lasted an hour or so, after which an operating list was drawn up to start at about 8pm when the nursing night staff came on duty.

Meanwhile, we would adjourn to the doctors' dining room for a free evening meal, often a juicy steak with a nice disc of garlic butter on top, served by waitresses in yellow dresses, with white aprons and caps. On Thursdays there was, in addition, a half-pint bottle of amber ale at each place setting, provided *gratis* by Scottish and Newcastle Breweries. Most doctors felt it unseemly to risk smelling of alcohol in the proximity of patients, so the beer was usually saved up for the occasional party in the 'house'. Such customs belong to a bygone age.

We then returned to observe any operations. A record was kept on Sister's desk of all admissions in chronological order, and the personal number of each student appended to their details, so that we had 'our' patients to clerk in, examine

---

1. Nowadays surgical specialties sensibly don't plan elective surgery when they are on call for emergencies. Operating theatres are left vacant but staffed for the timely treatment of such cases. In the 1960s and up until about the nineties, however, this was not common practice. Routine work continued as normal and only if an emergency was actually life-threatening was this pattern interrupted. In fact, many surgical emergencies in the category of 'acute abdomen' benefit from a short period of pain relief and fluid replacement. In practice, most cases dribbled in from early afternoon onwards and especially towards evening, mainly because they were not self-referred via the Accident Room but seen by their own GP, who tended to do house calls after morning and evening surgeries.

and follow throughout the duration of their admission. In surgery, this meant scrubbing up and assisting with the operation on our particular patients. The theatre sister showed us the technique of 'scrubbing up', a ritual that must be adhered to in order to preserve asepsis, and the method of donning gown and gloves without breaking the rules. We acted usually as second assistant, simply holding a retractor. We were often supernumerary in, say, an appendicectomy, but at least one of us obtained a good view. Other colleagues would simply observe, wearing a clean gown (not sterilised) over their outdoor clothes, plastic shoe covers and cloth hat and mask. Soon, disposable paper hats and masks appeared.

We were expected to stay until late and were usually questioned the following day about the cases. We could sleep overnight if we wished, but this was not compulsory. A small dormitory was provided and a porter woke us up if anything came in during the night.

'Gentlemen,' he would politely announce. 'Mr Lishman requests your company in Leazes Theatre, there is a perf.'

All very formal, old-fashioned and delightful.

Most of the 'medical' cases we saw in the wards were stable, either having investigations or undergoing treatment for their malady, but not acutely ill in the sense of being in immediate danger.

I witnessed my first miracle (at least, the first that was apparent to me) in the Accident Room medical cabin. The houseman was a slightly built young woman with short blonde hair, angular features, sensible flat shoes and an open white coat that seemed too big for her. The liquorice loop of her old-fashioned Sprague-Bowles stethoscope sprang out of one coat pocket with her BNF (British National Formulary) in the other. In the top pocket was her grey 'bleep' (the word 'pager' had not yet crossed the Atlantic), together with a Harry Hill row of pens. Her patient was an elderly man who was struggling with every breath. He was sweating profusely and had the staring eyes of a stag in the chase. I felt intrusive and fearful for him. Reassuring him gently with calm confidence, she slid into a vein a large syringe-full of Mersalyl. This was a wonder drug in its day but would shortly be obsolete, replaced by frusemide (Lasix), still the intravenous diuretic of choice. But Mersalyl did its job and within minutes his breathing eased and the crisis was over, for the present. He had acute pulmonary oedema (water on the lungs), caused by heart failure, and his prognosis was poor. Years later, I told this lady, Ann Marshall, who became an anaesthetist and with whom I did many cystoscopy lists, about my first vision of her. She laughed and blushed, but I think she was touched.

***

In November 1963 we had our first exposure to obstetrics (pregnancy, childbirth and the postpartum period). The junior course in this our fourth year was at Newcastle General Hospital. The three obstetricians were Linton Snaith, Hugh Arthur and Dorothea Kerslake, all in their early fifties. Linton Snaith was a very tall, thin man who, uniquely in my experience, wore white thigh-length rubber boots in theatre, like those worn by fish filleters. As befitted a gynaecologist, he wore a bow tie. He was formal in his treatment of students and a good teacher. Hugh Arthur was a quiet, kind man with a small moustache. He was a devout Scottish Presbyterian and an old-fashioned gentleman. Dorothea Kerslake was a character. Her hair was in the Eton-cropped style and she wore tweeds and brogues, with thick stockings. She was a chain-smoker, literally lighting one cigarette from another, and possessed a deep, masculine voice. She was a feminist, although I'm pretty sure she wouldn't have termed herself as such. If the girls were being reticent in seminars she would merely chide them with, 'Come on girls, don't let the side down.' She was notorious for carrying out abortions long before the procedure became legal in 1967. I have no idea how the potential patients were referred—whether by telephone or letters couched in euphemisms—nor whether she agreed to help married women or just those girls who had 'got into trouble', but she put them on the operating list as D&C (that is, dilatation and curettage), what the public used to call 'a scrape'. D&C was a common procedure in women with heavy or irregular periods. It was mainly diagnostic and, to some extent, therapeutic. Apparently, she always insisted on seeing the putative father and rumour was that she gave them such a bollocking that they were unlikely to make the same mistake again. She could not get away with such practice today—although, of course, her views became accepted and Parliament passed David Steel's Abortion Bill in 1967. Many people would say it has so many loopholes that abortion has become just another form of birth control, which was not the original intention, but that's another story.

The senior registrar on the unit was a dreadful man. I do not know how he achieved this position, nor why he wished to. He was arrogant, rude and appeared to be a misogynist. He once admonished a lady who was understandably shy about having a vaginal examination, especially in front of students.

'Come on, woman, open your legs,' he said. 'You're not holding your wind in at the vicar's tea party.'

He was universally disliked by students, especially the girls.

The chief midwife and deputy matron of the hospital was Miss Bradley, a stern woman in a green uniform. She ran a tight ship but she did show her human side at times. The Beatles were in their ascendancy and we would sing, 'You treat me badly, Sister Bradley, you've really got a hold on me.'

Students were in residence for one month and slept in a house opposite the General, called Lynwood. The midwifery department was at the end of a long south-north corridor of the hospital. During the day, in between ward rounds and Caesarean sections, we relaxed in a hut adjacent to the ward, waiting to be called to a delivery. We were assigned to women in labour in turn, and it was anticipated that we would deliver the baby under the supervision of a staff midwife. But it didn't always work out that way. We had a target to reach of twenty births. However, we were in competition with student midwives who were also present and we suspected that the staff favoured them over us. This was understandably so, for it was more important for them to gain the necessary experience.

It was almost unknown for a husband to sit with his wife during labour, and there remained an element of shame for an unmarried mother to give birth. The midwives kept a box of wedding rings in a drawer in case a girl wanted to wear one. Even a normal delivery could be dangerous in a woman with heart disease and we were still in an age when rheumatic heart disease, usually in the form of mitral stenosis, was not uncommon. These patients were looked after before and during labour by a much-loved cardiologist, Dr Paul Szekely, who qualified in Prague, and had an international reputation on the subject, publishing a classic textbook on heart disease and pregnancy, in collaboration with Linton Snaith.

The neonatal unit was beside the delivery suite and managed with neurotic military discipline by a Sister Gray. The paediatricians were Drs Fred Miller and Cyril Noble. In their presence, Sister Gray stood permanently fiddling with her rigidly starched cuffs, first one, then the other. Fred Miller insisted on students wearing meticulously clean white coats, which he maintained were our uniforms, and would dismiss anyone whose coat showed the slightest suspicion of a smear of dirt.

One day, the Indian paediatric registrar took a few of us to Dilston Hall, near Hexham, for a ward round. This small maternity unit served as an overflow for the city to the east. As was often the custom of the time, the two senior midwives in charge treated us to tea and cakes afterwards. In front of a roaring fire, these ladies entertained us with the history of Dilston Hall and Castle and how James Radclyffe, third Earl of Derwentwater, bade farewell to his young wife, Anna Maria, and rode out from his ancestral home in 1715 to join the Jacobite Rebellion. It's a well-known story to natives and lovers of Northumberland but was new to me and my colleagues. The rebellion was short-lived, the Jacobite army being soundly defeated at the Battle of Preston, and Derwentwater was beheaded for treason on Tower Hill in February 1716. His younger brother, Charles, escaped to France but was captured in 1745 on his return to support the later uprising and was executed in 1746. He was one of the last people in England to die by public beheading.

In a convincing manner, both ladies described one occasion at the anniversary of the Earl's departure when a horse was heard galloping past the hall in the night. The following morning they witnessed hoofprints in the snow leading to the Devil Water (or Devil's Water), the nearby stream that enters the Tyne just west of Corbridge. Mysteriously, there were no prints on the other side nor any sign of them returning. Our two midwives were quite excited that the story had recently been given prominence by the publication of the novel *Devil Water* by Anya Seton.

For some reason this episode, together with the name of the book and its author, remained in my memory. But it took another forty-five years and retirement before I read the book and revisited Dilston, which, incidentally, is also supposedly haunted by a grey lady. Anya Seton was an American author of historical romances, well known for her extensive research. Between 1941 and 1975 she produced twelve novels, two of which were made into Hollywood films. She had relatives in Felton in Northumberland and visited the county from an early age. Her father, Ernest Thompson Seton, was born in South Shields, but the family emigrated to Canada when he was a child. He later became an American citizen and was a well-known wildlife artist and author who had a major influence on Baden-Powell and founded the Boy Scouts of America. I found the book a fascinating read. The earl's estates were confiscated after his execution and the main complex was demolished in 1768, leaving only the remains of the castle, the chapel and gateway. The maternity hospital is now owned by Mencap. On further reading, I discovered that Radclyffe set out in October, which is a bit early for snow—disappointing, but never mind, it's a good tale. Perhaps the grey lady is poor Anna Maria, vainly looking out for her husband's return.

We were in our hut on November 22nd when we heard over the radio the news of President Kennedy's assassination. We were shocked like everyone else, but apart from the Cuban missile crisis of the previous year, I don't think any of us knew much about the Kennedys, except that they made an attractive couple, and that Jackie, for all her supposed intelligence and sophistication, had a silly, childish voice. Nevertheless, it was a terrible tragedy for that poor woman, not helped by the seemingly crass insensitivity of the apparent necessity for the new president to take the oath in the presence of a shocked, grieving widow, still spattered with the blood of her late husband.

November 26th was my twenty-first birthday and at 10.25am I delivered a male baby of 7lb 15oz to a lady of twenty-eight. It was her ninth (!) pregnancy, two of which had ended with miscarriages, and in one other the baby had died after four days, suffering from severe spina bifida. That evening a few of us went to the Prince of Wales pub beside the hospital on Westgate Road and got very drunk on 'stingers', a mixture of brandy and crème de menthe, a drink of which I have not partaken since, due to the resulting hangover.

It was an RVI tradition to hold a Christmas concert in the non-clinical staff canteen at the very end of the long main corridor, where there was a stage, curtain and lighting paraphernalia. The resident house staff collaborated with such nurses who were interested in writing sketches, acting, singing, dancing and playing musical instruments. It's a miracle that it appeared annually without fail, despite the housemen being on-call on at least alternate nights. It usually took the form of mickey-taking of the consultant staff and it was customary for the script to be vetted beforehand by Matron and the senior surgeon or physician. It was held for three nights and preceded by a dress rehearsal on a Sunday afternoon in front of ambulant patients, a receptive audience, despite probably not understanding the in-house jokes. It no doubt relieved their boredom and cheered them up. It was customary to hold a party after the final performance.

Meanwhile, my hospital training continued. Every lunchtime there was a lecture by a different clinician, some of them characters the like of which we don't see any more. One such was Dr Alan 'the Ogg' Ogilvie, a chest physician close to retirement. He was a tall, ungainly man with a long stride, wavy white hair, a wide mouth and thin lips. His physical awkwardness belied the fact that as a young man he was an athlete. He had a variety of both physical and verbal tics. Renowned for his eccentricity, he once entered the ward wearing one black and one brown shoe. When this was pointed out to him, he replied, 'Do you know, I've got another pair just like these at home!' Probably the most frequent story told about him is the time he performed a rectal examination on a patient during a reception ward round early one winter evening. With the patient lying curled up on his side, a gloved finger up his bottom and staff and students in attendance, the Ogg screwed up his eyes in concentration as he gazed at the window.

'Oh dear, oh dear...' said the Ogg.

One of his staff, thinking he was less than tactfully expressing his concern at having found a tumour of the rectum or prostate, asked what the problem was.

'Oh, I've just seen the new moon through glass and that's bad luck,' said the Ogg.

Until the advent of what are termed H2 receptor antagonists, such as the drugs cimetidine and ranitidine (trade names Tagamet and Zantac, respectively) there was no medical cure (as opposed to surgical) for duodenal ulcer, a common complaint. Symptoms could be relieved by a variety of antacids such as Milk of Magnesia, Aludrox and Rennies, and most sufferers learned to live with their condition, unless they developed a complication such as haemorrhage or a perforation. However, patients were commonly admitted with severe pain, which was difficult to control and referred to as 'acute exacerbation' of duodenal ulcer. One Christmas Eve, such a patient was admitted as an emergency under the Ogg, who saw him on his customary post-reception ward round the following day. He was being treated by a milk drip, a simple but effective method in which milk from a bottle suspended from a drip stand was continuously dripped into the stomach via a nasogastric tube. This was the only source of nutrition until the symptoms subsided. The Ogg looked at him.

'Oh dear, Sister—poor man,' he said. 'No turkey on Christmas Day? No turkey on Christmas Day? Oh dear! Sister—speed up his drip!'

In our introduction to surgery, Peter Dickinson, who later became one of my mentors, took a group of us to see a man with an inguinal (groin) hernia. He asked the patient what his problem was.

'Why, ah've gor a lump in mi lisk,' he replied, in broad Geordie.

Peter D's eyes lit up—he looked around and, perhaps unkindly, turned to our Norwegian colleague, Arne Walløe (pronounced Valloer).

'Mr Wall, what does he mean?' he asked.

'The name is Walløe,' Arne replied.

'Oh, I thought the 'o' was crossed out,' said Peter, again, unnecessarily unkindly.

In his undulating accent, Arne told him the man meant he had a lump in his groin.

'How did you know that?' asked a puzzled Peter D.

'Because *lisk* is Norwegian for groin,' Arne informed him.

Hoist on his own petard. Our first experience of the strong relationship between Geordie dialect words and Scandinavia.

1964 turned out to be one of the most important years of my life, because it was then that I met the girl who was to become my wife. I was invited to a party in Cedar Road, Fenham, so off we went in Niall's minivan. As the invitee, the others pushed me forward when the door was opened by a stranger to us, but who turned out to be Liz Rowell. I explained that 'we' had been invited by Vivienne, the red-headed staff nurse from ward twelve. In her usual forthright way, Liz asked if the one at the back was with us—it was Nick Wright, of course, he of the broad Somerset accent and way with the ladies, and she somewhat reluctantly allowed him in.

Some images of our past are imprinted like photographs in our memory, while others are visually vague and out of sequence. But my first sight of Mary Lack was in the kitchen at Cedar Road and I can remember it as if it was yesterday. She was in a beige jumper, had short, jet-black hair and cupid lips, but her most striking features were her large brown eyes. She was cheerfully chatting in a group, none of whom I knew, so she was unapproachable. I cannot truthfully say it was love at first sight because I was so lacking in confidence with the opposite sex that I felt she would be way out of my league. But the impression she made on me was profound. She was one of the four nurses in the flat. It turned out that Liz's reluctance to admit Nick was well-founded—I later discovered that not only was he sick in one of the girls' beds (I should stress that she was not in it at the time), but he placed a saucer on the ballcock of the toilet cistern as a practical joke, so they woke up to a flood. Fortunately, I was not deemed guilty by association this time. The sequence of events that followed shortly afterwards is vague, however. Suffice to say that I went out with Liz very briefly, and then, by some miracle, Mary and her friend Dorothy Wills agreed to partner me and Roger to the annual

Medical Ball. Not only that but Mary inherited her late father's Vauxhall Wyvern, and she picked us up and drove us there. I remember seeing the car coming down Myrtle Grove and saying to Roger that I must be on my best behaviour because this was one opportunity I did not want to mess up.

Mary was a staff nurse on the children's clinic but was due to embark on the midwifery course. I became besotted and went nearly every evening to visit her in her new room at the nurses' residence in Framlington Place. I visited so often that I felt I might be responsible for her failing her exams due to lack of revision time. Fortunately, she didn't.

Mary passed her driving test at seventeen, whereas I hadn't yet been behind a steering wheel. I always felt emasculated in the passenger seat. I admired the way she handled the gearstick, which in the Wyvern was on the steering column. To my eyes it looked complicated, and she looked very sexy in her short skirt and fluffy pink coat. I was sure she would eventually drop me. But I gradually relaxed and after about three months we were going steady and agreed to marry when I qualified.

The Vauxhall was not a permanent fixture. She couldn't afford to run it on a regular basis, but borrowed it occasionally. On a visit to Holy Island, I gallantly contributed to the petrol and made a packed lunch. Unfortunately, we chose to eat it a stone's throw away from a herring gull gulping away at a huge eel. The fish gradually disappeared down the bird's throat as we sat, mesmerised and a little nauseated, tentatively chewing our salmon sandwiches.

Chapter Nine

The Flying Squad

In the summer of 1964, those of us who had passed 2nd MB at the first attempt had to arrange a three-month elective period at home or abroad. This excluded my flatmates, Roger and Jim, who had failed. Several crossed the Atlantic to the USA or Canada. After all, one could get a flight for about £45 then. But this was half a term's living costs to me, so Niall and I went to Truro, in Cornwall, which sounded positively exotic compared to George's choice of Whitehaven, in Cumbria. However, because they could only take us for six weeks, we had to occupy the remaining time in pathology, which was no hardship considering that a year hence we would take the 3rd MB of pathology, bacteriology and public health. We assisted at autopsies and viewed slides and saw what the full-time job entailed. From the point of view of interest, I could have happily been a pathologist, but medicine is really about contact with live patients, coupled with the thrill of the unexpected.

We travelled to Truro in Niall's van, with an overnight stop at his prospective parents-in-law in Berkhamsted. The Royal Cornwall Infirmary (RCI) was a grey stone building that has since been replaced by a modern one of the same name. The elective specified that we did medicine, as opposed to surgery, and we were attached to a physician with an interest in diabetes. We were expected to attend ward rounds and clinics, but they anticipated that we would enjoy ourselves to some extent and not be totally tied to the hospital.

Most of the junior staff were from India, although one houseman was British, having qualified at Bristol, the 'feeding' medical school for the south-west before the creation of the new Peninsula Medical School at Exeter. One of the Indians, Acharya, was a character. His sense of the absurd always seemed in evidence and he was usually cheerful—well, except for the time he had to transfer a patient in a helicopter and returned as a nervous wreck, much to the amusement of his fellow countrymen. The operating theatre was directly above the residents' common room and every time the surgical diathermy was used during an operation it caused a temporary auditory and visual blizzard on the TV screen. Acharya gave a running commentary on the progress of the operation as perceived by the frequency of the disturbed transmission.

'He's stopping the bleeders in the subcutaneous fat. Ah! He must be opening the peritoneum now.'

He would follow this some time later by, 'He's just tidying up on the way out—he's closing now.'

I realised how isolated the south-west peninsula was. For example, we did an outpatient clinic at Newquay. Here the medical facilities were geared towards its winter resident population, even though for at least three months a year, this number rose significantly. Yet if you fell and broke an arm on the promenade on your holiday you had to travel thirteen miles to Truro for an x-ray!

The small hospital at Falmouth had the only circular ward I have ever seen, with a central nurses' station. It was very practical, as the patients were all easily visible. It was here that I saw an obese lady sitting up out of bed, her swollen legs resting on a red rubber sheet, with both feet in a tub. Several narrow steel tubes extruded from her legs, draining fluid into the receptacle. This was the only occasion in which I saw Souttar's tubes in use—the idea was to evacuate the fluid swelling (oedema) of congestive heart failure (or dropsy, to use an old-fashioned term). This would ultimately prove futile and any benefit was only fleeting, but at the time, digitalis and diuretics were the only therapeutic agents for this condition.

I got a pleasant surprise one morning when Roger arrived at the hospital. Dorothy and Mary had booked a week in a small hotel and Roger made his own way (by hitching, I think; he was always careful with his money, like a true Yorkshireman). He and I went to find him a room in a pub and so for the next week I made only a token presence each morning at the ward or clinic. We visited St Michael's Mount and Perranporth beach and spent every evening in the Thomas Daniell pub near the hospital. The only hitch came when we went to Mousehole and I tried to recapture part of my early youth by eating some mussels. As we left the shop I pointed to a sewage pipe leading out to sea and in a jocular aside, said I hoped they did not harvest the mussels from those encrusting it. But I think they did, for I got severe bloody diarrhoea, which persisted until I returned home to Hull after the elective. In typical fashion I never sought medical advice, having illogical thoughts about losing my colon and wearing an ileostomy bag for life—a little knowledge and all that. Fortunately, it eventually subsided.

Although our duties were confined to medicine, we did see the consultant surgeon in passing. There must have been more, but we only saw the one. He looked almost cadaveric—grey, sunken, lined cheeks and teeth which stood out because they were edged with tar from his chain-smoking. Nearly every night his sports car was parked outside until late. He, too, was probably suffering from the effects of a suddenly increased population. It should have been enough to deter me from a career in surgery, but rarely does youth look too far into the future.

'He'll kill himself one day, that man,' our physician said, meaning he was working too hard.

Sure enough, I saw his obituary in the British Medical Journal only two or three years later.

With no exams until the following summer, we could relax a little. Mary and Dorothy worked twelve-hour shifts, so we didn't see them as often as before. We often met in the Bun Room for a drink (or two). The physiotherapists held a dance in the fracture clinic, well away from the wards, one night. I was drinking in the Bun Room beforehand with Gerry McLoughlin, who has not yet featured in these pages. Gerry was and still is a good friend. He was extremely bright, very funny and unfailingly generous, but could at times exhibit those marginally psychopathic tendencies associated with the Celtic genome, especially when 'in drink'. We decided to leave for the dance and the Bun Room was packed, so instead of slowly making our way through the throng to the exit, we decided to climb through a window onto the grassy ridge outside the basement-level building. Gerry went first and told me to get out of the way as I followed him. When I did nothing because I didn't understand what he meant, he repeated, quite aggressively, for me to get out of the way, and I saw that he still had an empty pint glass in his hand. I edged aside and he hurled the glass with all his might into the room, where it struck the wall opposite and shattered into pieces. Well, it was no time to stand still and shrug one's shoulders, so we both took off for our lives.

At the entrance to the physios' dance was a sign that read, 'Owing to the proximity of the wards, anyone creating an unnecessary disturbance will be forcibly ejected.' I think Nick must also have been there because his flatmate, Big Jim, a very tall Scotsman who was reading geography, was, and he went nowhere without Nick. He spent the evening dancing with a similarly tall African lady, with a metal bedpan inserted down the back of his trousers (probably because he was a bit tipsy, found one, and thought it was amusing to do so!). Gerry started becoming loud, or aggressive, or both, and one of the snooty physios (they always felt they were a cut above the nurses) took him aside and pointed to the notice.

'Can't you read that?' she asked.

'Owing to the proximity of the wards,' McLoughlin slurred in reply, 'anyone creating an unnecessary disturbance will be forcibly ejaculated.'

At the start of Michaelmas Term 1964, the four of us moved to a different flat in Glenthorn Road. Shortly afterwards, Mary and Dorothy joined Anthea Heagney and Felicity 'Flick' Rowbotham in Mayfair Road, which was close to us, separated only by an underpass beneath a railway track. In the autumn I met Mary's mother at her home in Bishopton, a pretty village near Stockton-on-Tees, County Durham. Rhoda Lack lived with her youngest child, John, then aged seven, having lost her husband five years previously. The purpose of the visit was not just to meet Mrs Lack, but as a courtesy, to ask for the hand of her daughter in marriage. By prior arrangement, Mary left me alone with her mother for a discreet interval. Fortunately, she had no objection.

In November we had our senior obstetrics placement at the Princess Mary Maternity Hospital (PMMH) on the Great North Road. We were resident for one month, as in the previous year at Newcastle General, and once again occupied a dormitory at the rear of the building. The consultants at the PMMH were Professor JK 'Jake' Russell, Derek Tacchi, Frank Stabler, Denys Fairweather, William Hunter and David Miller, and the senior paediatrician was Gerald Neligan. Jake Russell was a stern Scot, a stickler for dress code, etiquette and respect of rank. For some reason the white coats of the senior staff had blue collars and Jake's possessed, in addition, three sergeant's stripes on the upper arms to indicate that he was the boss.

Frank Stabler was a short man who had a bald pate with white hair at the sides. Approaching retirement, he was a jovial chap, widely respected for his wide experience and lifelong dedication to the improvement of the care of women in pregnancy. I assisted him one Sunday morning to do a Caesarean section on a diabetic lady. Diabetics often had large babies and sometimes underwent elective sections at thirty-eight weeks, two weeks early, to ensure that the baby was not too big. On this occasion he removed the baby and placed it in front of him, holding it between his tummy and the operating table. The Indian paediatric registrar stood behind him expectantly, holding a green towel to receive the newborn and transfer it to a resuscitation trolley.

'Let him wait,' said Frank. 'The extra blood will do him good.'

Despite the pleas of the registrar, he waited until the umbilical cord ceased pulsating before cutting it and relinquishing the by now crying baby. This had apparently been his practice for years, but I recently saw a newspaper article suggesting it was a new 'discovery'.

Derek Tacchi was suave, handsome and florid-faced, with wavy blond hair and the measured walk of a man who was comfortable in his own skin. The nurses called him 'Ticky-Tacky'. He had a large private gynaecological practice and was the obstetrician of choice for the wives of the local GPs. George and I accompanied him once on a consultation in the RVI and were directed to the patient's bedside.

'Good morning, Mrs X,' he said. 'My name is Mr Tacchi. Dr Y has asked me to give my opinion about your condition. These are two of my henchmen, Mr Bone and Mr Clarke. They follow me around like dogs.'

David Miller was the youngest consultant, still in his thirties, a popular teacher and hard worker. I remember him entering the theatre in his pyjamas in the middle of the night to perform a Kielland's forceps rotation, a notoriously difficult and potentially dangerous manoeuvre. Sadly, he died of a rare form of cancer only a few years later.

I still have my book of case records, all neatly transcribed in fountain pen blue ink. My first delivery, on November 11th 1964, was a tragic one, and a new lesson in life for me. It was the second pregnancy of a married woman of twenty-two, whose antenatal care and labour were uncomplicated. The male baby was born in the early hours but lacked tone and had delayed spontaneous respirations. The child was also cyanosed and moaning, so the midwifery sister called Hans Steiner, the paediatric registrar, who intubated him and detected a widespread systolic heart murmur. Within an hour, the baby was dead. Soon after, I had to stitch the small laceration that the poor mother had sustained at the birth. Her husband, a merchant seaman, was at sea, and the following day she was sitting up in the postnatal ward to which they had transferred her, but she was the only one without a baby. In those days, babies spent a significant amount of time in the communal nursery to allow mothers to rest, but nevertheless, it seemed both tactless and cruel for her to be there. I suppose there was nowhere else. She was the same age as me and in my youth, naivety and embarrassment, I never mentioned the baby in our conversations. She gave sad smiles and superficially appeared untroubled, but it was heartbreaking to see her sitting alone. She returned home after five days. Nowadays, I suppose there's a good chance that the cardiac defect, whatever it was (I never found out the result of the post-mortem) would have been detected at a routine ultrasound scan and dealt with accordingly, if possible.

Once again, I delivered a baby on my birthday, my twenty-second. This was a twenty-year-old with her second pregnancy, booked in because of rhesus isoimmunisation. In this condition, a rhesus-negative blood group mother, pregnant by a rhesus-positive father, will have a rhesus-positive baby. During the birth of the first child, some blood cells of the baby enter the mother's circulation, usually at detachment of the placenta or afterbirth, and the mother responds by creating antibodies. At the second pregnancy, these antibodies cross the placenta into the baby's circulation and, depending on their concentration, destroy blood cells to a degree that can range from anaemia to jaundice and even foetal death while still in the womb. In this particular case, the baby was slightly jaundiced, with peach-coloured cheeks and, as was the custom, received an exchange transfusion of 490mls via the umbilical cord within twelve hours of birth. This procedure is no longer necessary since it became possible to destroy the rhesus positive cells at the time of transfer during the first pregnancy by giving such women an injection of anti-D antibodies.

One afternoon, a porter at the PMMH surprised me by saying a district midwife requested my attendance at a domiciliary delivery. I was handed a Gladstone bag, given the address, and told to go by bus. My destination was a flat within Rochester Dwellings, a complex of separate blocks in the poorer part of Walker, each block being labelled alphabetically by a large white letter painted on its side, say from A to D. A high wall encircled the whole, access being via a couple of arches. The dwellings had been purpose-built to house 'problem families'—no political correctness then—and it certainly had some features of a prison.

On entering the ground-floor-level apartment, my feet immediately stuck to the carpet. A toddler trotted around dressed only in a soiled vest. A large, fat, jolly red-faced midwife faced me, sitting in an armchair, side-on to a roaring fire. To my huge relief I was too late, and she was washing the newly born baby. As she chatted away cheerfully to me, the baby began to cry. Immediately the mother's voice emerged, unseen from the bedroom next door, and uttered, 'Get that bloody little trout out of here.' That was the welcome this poor little mite received into this strange new world, outside the comfort of the womb. The midwife rolled her eyes in despair and carried on drying the baby with gentle affection. What a start in life.

Just before dawn one cold morning, I was informed that the Flying Squad had been called and to wait at the entrance for the car, which was on its way to Blaydon, of races fame, where a woman with a retained placenta needed our attention. Professor Farquhar Murray founded the Obstetric Flying Squad, the first in the country, in Newcastle in 1935. Eventually, there were 160 such units nationally. The high number of women suffering from postpartum haemorrhages arriving at the hospital either dead or moribund had depressed Murray. He concluded that the journey was responsible and suggested a trial of taking a team out to the patient rather than vice versa, then transferring them after resuscitation. Of course, the 1930s was a time of great poverty, no stored blood for transfusion and no antibiotics. Furthermore, the majority of women were delivered at home and hospitals had a reputation for being dangerous places because puerperal sepsis due to the streptococcus bacterium was still common until the discovery of penicillin. The concept of such a Flying Squad was contentious but became accepted and ran successfully until about the early 1980s, when it became obsolete because so few births then occurred in the home. In the war years, Frank Stabler and Willie Hunter were the mainstays of the service, and it was the latter who picked me up.

Willie was not a very imposing figure, balding with a receding chin and a flimsy moustache, and despite his wide experience of the service, he belied its label of 'Flying' by driving his Rover 90 at a funereal pace. Despite the hour, he wore a three-piece suit, shirt and tie. The call came via a GP or midwife, and a relative, usually the husband, was instructed to wait by the roadside at some appropriate point with a white towel slung over the forearm for identification. He was then picked up and hence guided us to the address. Remember, this was long before satnav and even good street maps, and the catchment area extended deeply into Northumberland and Durham. They carried blood in a box, in the group of O-negative, the so-called universal donor, which can be safely given in emergency without precise cross-matching. A consultant anaesthetist was also in attendance.

In our case, the initial problem was not so much the retained placenta, as the patient was not actively bleeding, but the roaring fire in the bedroom grate, which was the source of oppressive, almost painful, heat. Fortunately, the antiseptic supplied in the kit had an aqueous rather than alcoholic base, and most of the cotton wool swabs were soaked in it to damp down the fire. The patient was then slung across the width of the bed and anaesthetised by Johnnie Wheldon. Willie must have been soaked through with sweat because he placed a long brown rubber apron over his waistcoat and trousers and scrubbed up as best he could before donning gloves and removing the placenta. I shall never forget the experience, but remember thinking there must be a better way to run a show. Nevertheless, I feel full of admiration for Willie and his ilk, who worked tirelessly to improve the lot of women and children, many of whom lived in pretty squalid conditions.

At this time we also had lectures and clinics in gynaecology. Stanley Way was a gynaecological oncologist at the Queen Elizabeth Hospital in Gateshead. He had an international reputation and was a real character. He performed radical operations for late-stage cancers of the vulva, cervix and uterus. This sometimes involved clearance of the pelvis, including removal of the bladder and rectum. Patients then needed their urine and faeces diverting, with two options. It could be into either a combination of an ileal bladder, in which the ureters are implanted into a short loop of small bowel brought to the surface, and a colostomy, which meant wearing two bags. The alternative was the dreaded wet colostomy, in which they implanted the ureters into the large bowel, requiring only the one bag but at the price of a stinking semi-liquid effluent, frequent urinary infections, and probably social ostracism. The latter was rarely performed, understandably. It was Stanley's practice, after such an operation, to carefully arrange the specimen on a green towel and photograph it with his own camera, because being across the river in Gateshead, he didn't have the services of the photographer in the Medical School. On one occasion shortly before Christmas, he sent a completed film to Kodak to be developed. The telltale yellow package was returned, secured with adhesive tape decorated with holly leaves and berries. Inside, together with his colour slides, was a note.

'Why not convert your slides into personalised Christmas cards?' it suggested, with the offer to do so.

Stanley said he was sorely tempted. The image of a vulva attached to a cervix, uterus and ovaries on a green towel background, with the message 'Happy Christmas from Stanley and Ruth' would have caused some interesting reactions.

Breast cancer was within the remit of the general surgeon but Stanley thought that as a gynaecologist, he should be able to treat the disease. Before the days of mammography, there were trials to detect breast cancer using thermography, as tumours tend to have a richer blood supply than surrounding normal tissue and would hence be warmer, the increased heat being shown colourimetrically on a picture produced by a machine. Because of his reputation, Stanley was asked to take part in a trial and a machine was provided. The subject stood before the detector with breasts exposed and, to enhance the contrast, a current of cool air was blown across. Stanley was always on the side of the poor and deprived and he bemoaned the fact that when the offer to be screened was circulated, instead of women from council estates coming forward, 'What did we get but the West Denton Young Wives Association, forty peroxide blondes, all on the pill and boobies shining like Belisha beacons!' (The contraceptive pill increases the blood supply to the breast).

***

On New Year's Eve, 1964, Mary and I got engaged. We went to H Samuel in King Edward Street, Hull, and bought a ring with a sapphire and a small diamond on either side. In January 1965 Winston Churchill died and Roger and I watched the funeral in the Students' Union. For our generation, he was regarded by many as the saviour of the country, but there were still plenty of men around from the time of the First World War who took a different view.

We did a short course in anaesthetics, supplemented by lectures from Professor Edgar Alexander Pask, a slightly built, quietly spoken man with a somewhat prognathic (protruding) jaw. Who would have known that this apparently in-significant-looking chap had made such major contributions to the RAF during the war? This contribution was at no small risk to himself but was not in combat.

He subjected himself to sessions in a decompression chamber to show that 35,000 feet was the greatest height at which a pilot might survive when escaping by parachute. He was anaesthetised many times to evaluate different methods of artificial respiration and he designed two flotation jackets, which ensured that an unconscious pilot would not be turned face-down in water. He tested them himself in a pool at Farnborough which could simulate waves, anaesthetised by his then-chief, Professor McIntosh of Oxford. Not only that, but he also designed clothing to protect subjects from hypothermia in water and was parachuted into freezing seas off Shetland to prove his concepts. He was only the second professor of anaesthetics in the country when he was elevated to that position at Newcastle in 1949. Of course, he never mentioned any of this. Such revelations only surfaced after his tragic early death at the age of fifty-four in 1966. He died from a massive coronary thrombosis, caused no doubt by heavy cigarette smoking, an unexpected habit for an anaesthetist.

A popular venue in our social lives was a visit to Balmbra's Music Hall in the Bigg Market. Pronounced Bambra's, it's mentioned in verse two of the Geordie national anthem, *The Blaydon Races*—

'We took the bus fra' Balmbra's, She was heavy laden, Aal the way doon Collinwood Street, That's on the way to Blaydon...'

For about 3s 6d one could be served drinks at bare wooden tables and be entertained by genuine back-end of the music hall tradition acts. The master of ceremonies, Dick Irwin, was a Geordie institution, decked out in top hat and tails in his lectern box at one corner, holding the gavel used to bring the house to order. He told virtually the same jokes at each show but, as they say, it's the way you tell them—a combination of a broad Geordie accent and timing, helped by the generous attitude of the audience. He would usually introduce himself by bragging that he had performed before royalty—the Prince of Wales, the Duke of Wellington, The Duke of York, and half a dozen other pubs down the Scotswood Road.

Later on, when I became a house physician, the widow of the proprietor was a patient. She had been a dancer herself but lost a leg in the Blitz on London and now had a deep vein thrombosis in her remaining one. Her late husband was in the music hall and his connections kept the tradition alive locally. She asked me if I remembered the man who collected our tickets at the door, which I did. She then explained that he had been with them for years and if an act was required at short notice, he could step in and do a spot of juggling. One night he performed the act perfectly, but when he left the stage walked straight into a door. His colleagues laughed because he appeared tipsy but was known to be strictly teetotal. Sadly, it was the first symptom of a brain tumour, which soon killed him. So that was the end of an era.

A few years later, when I was a registrar in urology, Dick Irwin was one of our patients and was as irrepressible as ever. Balmbra's still exists. It retains the name, but is simply a bar and a shadow of its former self.

Chapter Ten

Close Shaves

In March 1965 we took the 3rd MB exam in pathology, bacteriology and public health and as far as I can recall, we all passed. The next hurdle would be Part 1 of finals in December, in obs and gynae. We then had our first exposure to paediatrics, some of it in residence. The Department of Child Health, as it was called, had an excellent reputation deriving from a man we never knew, the charismatic Sir James Spence, who died in 1954. His legacy lived on, especially in the Babies Hospital in Leazes Terrace, next to Newcastle United's St James' Park stadium. Like our anaesthetics lecturer Professor Pask, Spence was only the second professor in his specialty in the country when appointed in 1942. In 1947 he began the Thousand Families Study, which became nationally important and was continued after his death by the Newcastle team led by Professor SDM Court.

They did not assign us to specific clinics and we could go where we liked. The most popular clinic was that of Ernest Brewis. He was easy-going, sensible and also specialised in infectious diseases, in adults as well as children, and he seemed more 'human' than some of the others. This latter side of him revealed itself in an obvious broken nose received while boxing in the Royal Navy. Rightly or wrongly, we felt that some paediatricians had become 'infantilised' by their prolonged exposure to children, in the same way some teachers are. I now feel guilty about having these thoughts because, in retrospect, it was a privilege to be associated with most of them, however peripherally. They were incredibly dedicated to improving the lot of children and several had deserved national reputations.

When we were resident in the RVI, there was a ward round every night at ten o'clock. I don't know why, because most of the children were asleep. The two children's wards, eight and sixteen, were decorated with tiled panels made by Royal Doulton of Lambeth and dated from the opening of the hospital in 1906. These were—and fortunately still are—of nursery rhyme characters such as Little Boy Blue, Simple Simon and Little Miss Muffet, and every tile is hand-painted. Several hospitals both in this country and overseas possess similar collections made at around the same time, but that of the RVI is the biggest in the world, with some sixty panels containing thousands of tiles.

One of the ward sisters was a severe-looking woman who rarely, if ever, smiled and seemed to me totally unsuited to her job. I shall not forget the poor boy with tuberculous meningitis who was subjected to daily injections of streptomycin in the spine via lumbar puncture for thirty days. Each morning she took him to the treatment room and clasped him across her knees with one arm behind his neck and the other below his buttocks so that he was in the correct position, with spine curved towards the operator. The poor kid yelled but he was immobile, clamped as if in a vice. Talk about being cruel to be kind.

I was resident at Walkergate Hospital, the venue for infectious diseases, in the east end of the city, and the fiefdom of Dr Brewis. The commonest admissions were for infective diarrhoea, in those days usually salmonella or shigella. The ward sister had worked there for years and it was said that she could identify the bacterium responsible solely from the smell in the ambulance.

Although I was doing paediatrics, I also saw several interesting adult cases. One Sunday a man was admitted as an emergency, almost dead from dehydration. In the past he had required the removal of the whole of his large bowel for chronic ulcerative colitis and he now had a permanent ileostomy. This is where the end of the small bowel is brought to the skin surface like a spout and the semi-liquid motion empties into an attached disposable bag (although in those days the bag was of rubber and had to be washed out). He was wildfowling in north Northumberland at dawn when he stumbled on a duck's nest containing eggs. He took one, cracked it and swallowed it raw—he must have been peckish. Within a very short time, fluid began to pour from his ileostomy and by the time we saw him he was moribund and almost unresponsive. The ileostomy was working constantly, clear fluid almost spurting out like a mini-fountain. Fortunately, he soon recovered with fluid resuscitation, but this episode taught me that if you want to eat duck eggs, make sure to cook them properly. Ducks frequently harbour salmonella in their oviducts, which contaminates the eggs.

I saw an adult with chickenpox for the first time. He was about thirty-five and caught it from his daughter. The poor man felt a complete fool and was thoroughly miserable. She had quickly shaken it off as children do, but he was completely covered in the rash, needed to wear dark glasses and had pneumonia and a deep vein thrombosis in one leg.

A shocking and depressing experience was seeing the small unit where victims of bulbar poliomyelitis (also known as infantile paralysis) were nursed. Suffering the severest type, in which the nerve supply to the diaphragm and intercostal muscles is damaged, they were permanently in iron lungs and were here to provide a couple of weeks' respite for their carers, usually relatives. They were disembodied heads, the only part of their bodies visible, protruding from the rectangular box which encased them and providing both their respiration and access for the carer to perform ablutions. What a miserable existence! Thank God that during my lifetime this scourge of mankind—and especially children—has been virtually eradicated, apart from scattered pockets in Pakistan and the former USSR. I had the privilege of seeing the creator of the oral vaccine, Albert Sabin, lecture to our medical society. It's a pity he didn't live to see the results of the fantastic World Health Organisation vaccination programme.

In the evenings when we weren't on the wards we could relax in a common room in the residence. It was there I met a character by the name of Beresford Hall-Parker. He was difficult to age, but certainly at least in his forties. Married with five children, he was the ophthalmology SHO. Most SHOs were usually in their twenties, but Beresford was a man who made things difficult for himself by adhering to his principles. He had an unruly shock of greying hair, a large moustache and wore scruffy tweeds. His jacket pockets hung down well below the cloth due to the weight of his pipe-smoking paraphernalia. He monopolised conversation but was entertaining and quite content to continue past midnight. It seemed rude to stop him in mid-flow and retire to bed.

The conflict with his principles dated as far back as his application to university. He had to obey the rules of passing maths, English and a foreign language at matriculation (the forerunner of O-levels) as well as the equivalent of A-levels. According to Beresford, he refused to sit French because he claimed it had nothing to do with medicine. This stubbornness precluded his admission to University College London and he had no alternative but to take the LMSSA (Licentiate in Medicine and Surgery of the Society of Apothecaries). This has been termed 'the backdoor to enter medicine' as it was devised for those candidates, often from overseas, who didn't take the usual route through medical school. There has been some dispute as to whether the examination was as tough as university finals, but this apart, it begs the question of how candidates fulfilled the requirements of the long medical course. Beresford maintained that University College Hospital permitted him to attend classes but not sit the exams. All in all, it seems that it would have been easier to take an O-level in French.

His conversation centred around his bitterness at how this 'lower' qualification had damaged and delayed his progress. Despite his experience, he kept being overlooked for promotion because he didn't have the FRCS Diploma in Eyes. In fact it took him several attempts to gain even the Dip Opth, never mind the Fellowship, because he argued with the examiners. He quoted as an example being asked to look in a subject's eye with the ophthalmoscope and describe the pathology. On correctly diagnosing hypertensive retinopathy, he was then asked to grade it. Unfortunately, he replied that he did not use a grade, but simply described what he saw. Once again, it would have been easier to abide by standard practice and simply do what he was asked. The final straw was when an outsider was appointed to the next senior registrar post, and on his first independent operating session, the head of department instructed Beresford to sneak into the operating theatre to make sure the new man knew what he was doing!

When interviewed for his present post, he was asked if he owned a car. He told me he didn't wish to be dishonest and so answered in the affirmative. He omitted to explain that he couldn't afford to insure it and that it was propped up in the garage without its four wheels. He explained, quite bitterly, that a British appointee was expected to use his own mode of transport for the several peripheral hospitals it was necessary to visit, while they permitted those from overseas to use taxis at the department's expense.

I describe this at some length because Beresford did finally become a consultant ophthalmologist, at the North Riding Infirmary in Middlesbrough, the town where I, too, ended up. Eccentric to the last, he wore the apparel of a Harley Street specialist, morning coat with waistcoat and striped trousers, watch and chain—a little over the top for the Boro. His letters to GPs were short and to the point, but terminated with his full signature—Beresford Hall-Parker—in black fountain pen, which he had obviously practised to perfection so that it was all whirls and flourishes, reminiscent of that of Good Queen Bess in a proclamation. In fact, even while still in Newcastle he developed a good reputation outside his own department, and Professor Reg Hall, an internationally renowned endocrinologist, sent all his patients with the ocular complications of thyroid disease to see Beresford.

In the summer of 1965, one of Mary's flatmates, Anthea Heagney, asked several of us to a party at her home in Guisborough, North Yorkshire, to celebrate her brother Michael's twenty-first birthday. Anthea was a quiet, undemonstrative girl, but good fun, and her long-time companion, I hesitate to refer to him as a boyfriend, was GH de G 'Hartley' Hanson, a budding ENT surgeon turned bacteriologist, who was a few years older than the rest of us. Hartley was bald, short, rotund, a devoted beer drinker and a member as we all were of the Medical Institute, a private drinking club, and the Royal Naval Volunteer Reserve. Based at *HMS Calliope*, which was now entirely shore-based on the Tyne, it was once a real ship, allegedly kept afloat by submerged empty gin bottles. He was also a great chap.

Anthea was a private girl, but was open about her devout Catholicism. Those of us invited borrowed a car and made the journey, and what a surprise awaited us. The large detached house, aptly named Tudor Croft because of its design, was in extensive grounds, with a stream and stone humpbacked bridge and a tropical plant house. Anthea's family warmly greeted us in a wooden panelled reception hall with stained-glass windows. After drinks and canapés, they bussed us into Stockton-on-Tees to the Fiesta nightclub, where the cabaret acts were of the highest quality. Only subsequently did we learn that this venue was one of the most famous provincial clubs in the country, often featuring national and international celebrities. We also discovered that Anthea's well-concealed family wealth derived from a chain of general stores in the Middlesbrough area.

On our return there was a barbecue and drinks galore, with a traditional jazz band dressed in striped blazers, straw boaters and cream slacks playing by the side of the outdoor swimming pool. An outdoor pool in the north? Unheard of! It was like something out of *The Great Gatsby*. Anthea must have had a quiet smile to herself; none of us expected anything like this.

One of our company, Graham Teasdale, a couple of years ahead of us, caused some consternation by diving into the shallow end of the pool. Fortunately for him, and for medical science, he only sustained a cut forehead. Graham studied neurology before switching to neurosurgery in Glasgow, and together with Professor Brian Jennett, helped to develop the Glasgow Coma Scale, now used throughout the world. It's as well he didn't end up with a coma himself!

As a finale to this episode, Anthea did not marry Hartley, and now lives in Dublin with her husband and six children. Michael, for whom the party was held, still lives at Tudor Croft, and opens the gardens annually to the public for charitable purposes.

We now started our senior surgical placement on the Professorial Unit, wards one and seven, with Professor AGR Lowdon, who was also Dean of Medicine. The Reader was Frank Walker, a sallow-skinned Glaswegian, with smoothed black hair and a wicked sense of humour. He was well-known for an interest in ulcerative colitis, as evidenced by his books on the subject, but at this stage in his career was also known for the length of time his operations took.

'What's the time?' said Frank during an operation.

'It's not a clock you want, Frank, it's a bloody calendar,' came the reply from his fellow Scot, Bob Pringle.

In later years, Frank was known for his speed.

Because the professor had many administrative duties, he had a first assistant to cover clinical work in his absence. In this post was L Brian Fleming, aka 'Flash', partly because of his operating speed, but probably also on account of his peacock-like fashion sense; Chelsea boots, narrow ties, rounded detachable white collars with striped shirts. There was also a Professor of Surgical Science, Denis Walder, who fortunately seldom operated and whose main research interest was Caisson disease (decompression syndrome) and Raynaud's disease. He was able to study the former extensively a few years later during the building of the Tyne Tunnel.

Professor Lowdon had a lot on his plate but took his teaching seriously. He once asked an Irishman how he was faring.

'Oi'm fadin' away loike the stars in the mornin',' he replied (he had cancer).

This was the one and only time I have heard this poetic expression and I seem to remember that its sad beauty stunned us into silence.

Lowdon was the first to tell us jokingly to always distinguish the Geordie pronunciation of work (walk), from walk (waak), exemplified by the question 'Can you walk?' to receive the astonished reply 'Walk? I canna even bloody waak!' translated as 'Work? I can't even bloody walk!'

One interesting patient was an old man in a cubicle. He was pink-faced with a bulbous nose and had a colostomy. He had intercostal pain due to osteoporosis of the ribs, for which he was to have nerve blocks by Dr Armstrong Davison, one of the senior anaesthetists. As students, we had a rota for blood-letting to help the houseman with his routine work. One of our number, Steve Craddock, a very serious sort of chap, simply picked up the blood forms and bed numbers and marched into the old man's room.

'Just a blood sample, sir.'

'What's it for?' said the old man.

'Just routine,' said Steve.

'What's it for?' repeated the old man.

'Well, if you must know, it's for your urea and electrolytes.'

'My urea and electrolytes are perfectly all right, thank you. Now, get out!'

Slightly taken aback, Steve checked the details, to discover that his intended victim was Professor Frederick Pybus, Emeritus Professor of Surgery.

FC Pybus (1883-1975) had qualified in 1906, the year the RVI was founded to replace the old infirmary on the Forth Banks, and committed almost his whole professional life to the hospital. Nicknamed 'Piggy', he was the last professor of the old school—that is, before the creation of the NHS. As a bachelor and a bibliophile, he did something impossible to reproduce today. With no family to raise and a good but not excessive income, over a period of thirty years or so he accumulated a large collection of antiquarian medical books. Finally, in 1965, despite enticements from this country and abroad, he generously donated what amounted to around two thousand volumes to the university library, where it is known as the Pybus Collection. Among them are a 1555 edition of *De Humani Corporis* by Vesalius and a 1648 *Exercitatio Anatomica de Motu Cordis* by William Harvey, autographed by his brother. In addition to the books are more than a thousand engravings.

In later life Pybus became obsessed by atmospheric pollution, probably prompted by his interest in cancer of the lung, and he wrote frequent missives to newspapers on the subject. He lived in a house called White Knights in Two Ball Lonnen, close to the RVI, and somewhat embarrassingly for his alma mater, aimed verbal slingshots, printed in the local paper, at the emissions from the hospital's incinerator chimney, which were visible from his residence.

Returning to the reason for his admission, Armstrong Davison was an old friend and one of the few whom Pybus would trust, which was a pity, because the former wore spectacles with very thick lenses and could hardly see to perform the required series of injections.

Keith Yeates, the urologist, worked for Pybus in the early 1940s, and told the following tale. The prof held a clinic on a Saturday morning, followed by a ward round, and on this particular day, the retinue had just reached the ward when the breathless registrar caught up with them.

'Professor Pybus, Professor Pybus, it's the last patient in the clinic, sir.'

'Well, what of it?' said Pybus, rounding on his heel.

'Well, sir—he—he wants a vasectomy!'

Today, one might wonder what the fuss was about, but then such a request provoked a response typified by a Bateman cartoon—that is, the retinue fall back in open-mouthed horror, while Piggy himself remains impassive, simply looks down his nose and enquires, 'Where does he come from?'

'I don't know, sir, does it matter?'

'Of course it matters, go and find out.'

The young man shakes his head and makes his way back to the clinic, muttering to himself that the old man must be losing his way.

(While he is gone, I will tell you why the request provoked the response it did. In those days, a vasectomy was only performed as an adjunct to the old-fashioned prostatectomy operation to prevent ascending infection from the bed of the prostate gland along the *vas deferens* to the testicles, which in the pre-antibiotic era could be pretty nasty. Unilateral vasectomy had been tried in the 1920s to 'rejuvenate' the recipient; the logic being that if the sperm-producing cells died as a result of the vasectomy, the remaining testosterone-producing cells might enlarge or increase in number, thus raising the blood level of that hormone. There was a vogue for the procedure in dodgy private clinics for Middle-Eastern potentates (or should I say, *impotentates*?) and there was the well-known case of the Irish poet WB Yeats, who had it done in 1934 and subsequently took up with a girl of twenty-six when he was sixty-nine (*post hoc ergo propter hoc*). However, as a means of sterilisation it was viewed as unethical, if not actually illegal.)

Our registrar finally returns to announce, 'If it's that important, sir, he comes from Gateshead'.

'I'll sterilise any bugger from Gateshead,' says Piggy. 'Put him on the list.' And he turns on his heel.

Another story about Pybus would seem unbelievable if I had not heard it from an eyewitness, Mr W Corbett Barnsley, the thoracic surgeon at Shotley Bridge. Pybus had a patient with a syphilitic aneurysm of the thoracic aorta, which was bulging through the sternum. The only diagnostic aid was an x-ray and for some reason the prof got into his head the notion that the aneurysm was saccular, rather than the usual fusiform—that is, a spherical sac coming off an otherwise fairly normal aorta, which was on the point of bursting and causing immediate death, but had a short narrow neck which could be ligated (tied off). Almost always, such aneurysms involved the whole arch of the aorta and were inoperable at the time. Pybus announced that he would operate shortly before leaving for a conference in Paris. According to Corbett, news of this risky venture spread like wildfire and the theatre was as packed as it could be with onlookers. Almost with his first incision, blood shot up like a fountain.

'Tonsillar swabs and lots of 'em,' Pybus requested, apparently unfazed.

He crammed in as many as he could under compression from wide silk stitches to staunch the bleeding and then calmly peeled off his gloves.

'Sister, I shall dress the wound on my return from Paris,' he coolly announced, knowing full well that the patient would shortly succumb, which he duly did.

On the male surgical ward, ward one, was an orderly called Jack, a very thin, middle-aged man with a pencil moustache and who wore a short white jacket. I suppose Jack's job entailed a variety of tasks but I remember him mainly for both shaving the men and his afternoon call of, 'Any eggs?' as he walked up and down the ward and the adjacent glass-covered balcony which looked out onto the entrance to the hospital. If any relative brought in eggs, Jack would hard boil them and make sandwiches for snacks.

He shaved those patients who were unable to manage themselves, but he also did the pre-operative shaves of the abdomen and pubic area. Watching him perform the latter was a treat. After lathering the area with his brush and soap he would unsheathe his cut-throat razor in dramatic fashion. The whole procedure was contrived. Holding the razor in the right hand with his little finger extended, he would lift up the penis by the tip between finger and thumb of his left hand, the remaining three fingers splayed out. Accompanied by a tuneless whistle, he would then do the business, while the patient either closed his eyes or, more usually, stared down in horror, facial muscles tensed, at Jack's deft strokes of the razor around the base of his manhood. Once the procedure was completed, he was gently rinsed and talcum powdered like a baby.

It was customary for students to attend and, if possible, assist at the operation of any patient assigned to them. One Thursday morning a patient of mine, a woman in her early thirties, was first on Professor Lowdon's list for a total thyroidectomy on account of papillary carcinoma. The prof greeted me in the changing room and briefly discussed the condition and how it was a low-grade tumour and that surgery would be curative. He then exchanged a few pleasantries with Jimmy, the theatre porter, explaining that he could only do one operation before leaving for a meeting of the hospital governors. He was assisted as usual by Flash Fleming, who, after the prof's departure, completed the day-long operating list himself. Flash had no junior surgeon to call upon, so he asked if I would help out for the rest of the day. I was glad to do so because it was interesting and made me feel important, and any criticism he aimed at me for clumsy or tardy assistance was rapidly curtailed by the theatre sister. She protested that I was only a student and, in any case, was acting voluntarily.

Thursday was reception day for the unit, so immediately after completing the list I went to the seminar room to join my colleagues, where we were met by Mike Kestle, the registrar, who told us that Prof Lowdon had died suddenly that afternoon. After the governors' meeting, he went for a walk near Blanchland and collapsed. He was only fifty-four. The shock was such that I found it almost unbelievable that only a few hours earlier I had been speaking to him. There was no hint throughout the day that anything untoward had happened. I suspect they kept the news from Flash because he was the prof's first assistant and close to him. I don't think any of us felt like being taught, but Mike Kestle continued nevertheless. It was September 2nd 1965.

It was only after his death that we learned he spent the six years of the war in the army, initially in Palestine and then with Montgomery's Eighth Army from El Alamein to Sicily, and later on the Normandy beachhead. While in Egypt he treated King Farouk for a scalp wound sustained in a car accident and was decorated with the Order of Ismail, 'Fourth Class', much to his amusement.

Andrew Lowdon's sudden death was obviously a grievous blow to his wife and their four children, but also to the medical school and university. Like Pask, and two close colleagues of my own in later life, John Farndon and Bill Leen, he died during what we in medicine refer to as 'the dangerous fifties', from coronary heart disease, before these talented individuals could complete the tasks the foundations of which they had created. Shortly afterwards I did a student locum for a week during one houseman's absence on holiday. As you might expect, the atmosphere on the wards was funereal.

In time a successor had to be found, both for a Professor of Surgery and the Dean of Medicine. As for the former, the Reader, Frank Walker, must have thought he was a shoo-in for the role, especially as he had gained his present post above his now main rival for the chair, Ivan Johnston, a Belfast graduate presently at the Royal Postgraduate Medical School at the Hammersmith Hospital in London. But it was not to be. In 1966, Professor Ivan DA Johnston, at well over six feet and with a voice to match, blew in like a whirlwind. Frank didn't receive the news well, and many years later Ivan told me that in the three months or so that they 'worked together', Frank did not speak to him once. In a fit of pique, Frank applied for the first consultant post he could, and it was at Hemlington Hospital, just outside Middlesbrough, a single-storey EMS (Emergency Services Hospital).

Although Frank had many admirable qualities, and I liked him, in my opinion he would have made a terrible Professor of Surgery. He took instant dislikes to certain individuals, and there appeared to be nothing they could do to redeem themselves. Like so many surgeons, he fossilised as he grew older, although this may have been due to lack of stimulus and might not have occurred had he been in an academic environment.

Professor Lowdon was succeeded as Dean by Dr Henry George Miller, the neurologist and charismatic extrovert, who became a Newcastle institution, especially during the years from 1968 when he served as university vice-chancellor until his untimely death at sixty-three in 1976. Henry Miller had a cheerful, handsome visage, but as a *bon viveur* he was grossly overweight. He was renowned for his energy, remarkable intuitive clinical ability, and his command of language, which often included outlandish remarks and barbed wit. He always dressed immaculately in black jacket with floral buttonhole and striped trousers, and if he swept by you he left behind the combined smell of *eau de cologne* and gin. Any lecture or clinical presentation by him was packed to the rafters in anticipation of a sparkling, humorous performance and he never disappointed us.

The other great Newcastle neurologist, John Walton, later Sir John, finally Lord Walton of Detchant (a hamlet in Northumberland), who remained hearty and active until his death at ninety-three in 2016, was about ten years younger than Henry. He was based at Newcastle General as opposed to Henry at the RVI, so the two were not quite contemporaries, but they had a friendly rivalry. In his autobiography, *The Spice of Life*, and in his obituary to Henry, Lord Walton writes generously about his personality and attributes but does not shy from commenting on his at times unpopularity among some of his peers, usually due to tactlessness.

I witnessed such an example when Henry introduced the new Professor of Anatomy, John Owen, in 1974. It was before the customary public lecture given by the incumbent, which followed the resignation of Raymond Scothorne, who had returned to Glasgow as Regius Professor. It has to be accepted that the main purpose of an anatomy department is to teach gross anatomy to medical students, but of course, this subject no longer lends itself to research. Research was usually at the microscopic level, involving embryology or the immune system. Scothorne had contributed significantly to the latter, and the senior lecturer, Doug Hally, studied the abstruse topic of the nasal salt glands of ducks. It was actually an interesting one, explaining how some birds can exist at sea by secreting away excess salt via such glands and surviving by drinking seawater. It was known that the new professor was keen on research into the immune response, but equally that his knowledge of gross anatomy was confined to that learned as a student. In his usual way, Henry pointed out that it puzzled him to know why the new man was so titled, because he didn't know any anatomy! I well remember thinking that I had misheard him, but I hadn't, and it was clear that the recipient of the barb was, understandably, not best pleased.

The combined publications and personalities of these two neurologists gained Newcastle an international reputation in the specialty. It seems invidious to compare them, especially as I knew neither well, but from a distance, it seemed that whereas Walton's rise was due to a combination of talent and persistent hard work, Henry's was effortless.

Chapter Eleven

Getting the General Idea

C onsidering that the majority of medical practitioners became GPs, we were exposed in those days to remarkably little of the specialty—and despite the prefix 'general', it is a specialty. This major defect was partially corrected in the year below ours, who embarked on a new curriculum based on greater integration and exposure to patients almost from the outset. Curricula have been continuously modified since, and at the present time, GPs have their own training scheme on similar lines to the hospital specialties.

We were sent to three specially selected practices, at Bedlington, Rothbury and Whickham. It is perhaps notable that none were based in central Newcastle and I can only guess that the surgeries were too packed to allow any meaningful teaching, but there may have been other reasons. Bedlington and Rothbury left an impression on me. Dr John Brown of Bedlington had been a GP for forty years when I visited him. It was clear that he was a pillar of the community of that small mining town. Although he wore a brown trilby, as we walked together down the main street his headgear spent more time raised in the air than it did on his head, as he doffed it first to one side and then to the other. 'Morning, Mrs Smith,' 'Morning Mrs Jones' etc.

He involved himself in virtually every aspect of the communal life of the town and for years had attended accidents down the pit itself. He seemed to typify the best aspects of general practice as described by AJ Cronin. I accompanied him to visit a patient on a new council estate and after the consultation, he took great pride in showing me around the house, as if it was his own and, as far as I recollect, without permission of the housewife, who simply looked on with a pleasant, amused expression on her face. It seemed a bit cheeky, but both she and I realised that he was demonstrating how much the living conditions of the mining community were improving compared to those experienced in former years.

The bus trip to Rothbury was an eye-opener for me. Without personal transport, I had seen little of the county and was struck by the beauty of the countryside and the delightful town, situated on the edge of the Simonside hills. The senior partner was Rex Armstrong, grandfather of the television presenter Alexander, whose father, Angus, eventually took it over. It was Rex's partner who took me under his wing for the day. He was a pleasant, interesting man. I arrived just as he was starting his house calls. The first visit was to the cottage of a retired gamekeeper and his wife. I can't remember which of the two was the patient, but once the wife got talking there was no stopping her, as she gave an excellent demonstration of the Northumbrian burr. At one point the GP caught my eye and signalled to look at the husband, who had just surreptitiously disconnected his hearing aid while he relaxed and puffed his pipe.

The next one could easily have been in an episode of Dr Finlay's casebook. We went to a large, detached house in open country, and let ourselves in. The doctor called up the stairs to announce himself and in a pink and white bedroom lay a lady, probably in her sixties, sitting up in bed with a white crocheted shawl around her shoulders. On a lambs' wool rug beside the bed sat a totally immobile, but live, white poodle. The lady had suffered a melaena—that is, an internal gastric bleed, manifested by the passage of tarry black stools. The commonest cause of this was a peptic ulcer of the stomach or duodenum, but there are others, including stomach cancer. I suppose the doctor would later have arranged investigations in the form of a barium meal, but at the time he was simply monitoring her progress by visiting daily and measuring her blood count to make sure she wasn't continuing to bleed. To do this he used an old-fashioned haemoglobinometer, in which a pinprick of blood from a finger was dissolved in one chamber of the instrument, which showed up green when held up to the light. The green of the accompanying chamber was matched up and when the shades were equal, the haemoglobin level was read off as a percentage of normal. This seemed like good practice, but nowadays virtually every case of melaena would be admitted to hospital for urgent investigation. The condition can be unpredictable and the patient can suddenly collapse; dangerous when living alone. I can only assume she had a housekeeper or cook because she couldn't do much for herself while in bed, but there was no one else around when we were there.

The GP told me that until only recently they had conducted post-mortems in the policeman's garage. Those were the days!

Lunch was also an experience. Back at the house, surrounded by oil paintings, we sat with a few others around a large dining table, on which was silverware, and were served cauliflower cheese by a uniformed domestic servant. Very different from the doctor's surgery I visited as a locum in the mining village of Burradon a few years later—it was a green railway wagon still on its wheels and one gained access via a step ladder.

***

As well as the main subjects of general medicine, surgery, obs and gynae and paediatrics, we also had exposure to ENT (ear, nose, and throat, or more long-windedly otorhinolaryngology), eyes (ophthalmology), psychiatry, dermatology, radiology (x-rays) and radiotherapy, industrial health, orthopaedics and venereology. We even had a one-week course in dentistry! Among the few times I saw a lecture theatre full to capacity or overflowing were when we had forensic pathology and venereology. Lectures in the latter were by Dr Macfarlane, an Edinburgh man. The two main venereal diseases were syphilis and gonorrhoea, with non-specific urethritis well behind. Chlamydia and herpes were never mentioned. Syphilis was still a significant problem, but fading fast, although it was still routine to include the WR (Wassermann reaction) in blood tests of every hospitalised patient. The disease was deemed the great mimicker, as in the third and final stage it could affect any part of the body and resemble a wide variety of unrelated illnesses. When it affected the nervous system it was still termed GPI (General Paralysis of the Insane), not very politically correct by modern standards. It was characterised by an odd high-stepping gait caused in part by the inability to feel the ground due to numbness of the feet. Aneurysms of the aorta in the chest were also a consequence, together with incompetence of the aortic heart valve, and in the worst cases they could erode the breastbone and push their way forward before rupturing. I never saw the latter except in photographs in old surgery textbooks I acquired over the years. For hundreds of years, both syphilis and gonorrhoea were grouped together as 'the great pox', or simply 'the pox', as opposed to smallpox. They were only separated when the respective causative bacteria were discovered in the nineteenth century. Manifestations of both are different, but confusion probably arose because some victims suffered from both simultaneously.

It is said that the great surgeon and comparative anatomist, John Hunter, inoculated himself on the glans penis with venereal pus from a patient to see what transpired. This probable myth, in my opinion, was promulgated by a London surgeon, D'Arcy Power, in his introduction to a Hunterian lecture at the Royal College of Surgeons many years ago and has been passed down through generations as an example of Hunter's selflessness in the pursuit of science. But Hunter was no fool, and if one examines the original source he describes 'the subject' being inoculated with pus. He was honorary surgeon to the old St George's Hospital on Hyde Park Corner in the late eighteenth century and would have had many indigent patients on whom he could experiment. I doubt very much if he would have used himself as the subject.

Gonorrhoea, or 'the clap', was the commonest venereal disease, characterised in the male by severe burning pain on passing urine, coupled with a yellow discharge. Until condoms became available it was an inevitable consequence, sooner or later, of visits to a lady of the night. Samuel Johnson's biographer, James Boswell, was a frequent partaker of the delights of Venus and was said to have had attacks of the clap numbering in the teens.

In the pre-antibiotic age, a frequent sequela of an attack was a stricture of the urethra, rendering micturition (passing urine) increasingly more difficult and painful, ultimately requiring dilatation with an instrument called a bougie. As a boy in Hull, if one visited a public toilet, there would inevitably be a large notice above the urinals with the message—

'VD' (in big red letters) 'Go to Mill Street Clinic, in Confidence.'

This was situated down a quiet side entrance to the old infirmary. The meaning was lost on me for a long time, but the rumour was that if one was unlucky enough to 'catch' VD, whatever that was, a probe with spikes attached would be passed up the penis and then pulled out. This urban myth is, or at least was, widespread, and I suspect it derives from either the exaggerated description of a urethral swab being taken (relatively painless), or the passage of a bougie for stricture. The bougie is metal, smooth, and passed after anaesthesia with a jelly—again, relatively painless when done skilfully (see later during my year as Mr Feggetter's SHO in 1968-69). More likely, however, is that it dates from wartime and/or National Service and was the warning used by the Regimental Sergeant Major to deter his new young charges from taking the risk.

When syphilis affects the skin in the later stages it can cause large ulcers called gummas, beneath which the bone is often destroyed. Dr Macfarlane showed a slide of an old man wearing a cloth cap and grinning inanely at the camera. In the next slide, the cap had been removed to reveal the total absence of the vault of the skull, exposing what looked like a very large shrivelled walnut. This was the brain covered by its desiccated membranes, the outer of which, the dura mater, is like parchment. It beggared belief that anyone could walk around, apparently untroubled, in this state. Similarly, there was a slide of a miner who had a gumma the size of a dinner plate on his back, and yet he continued to work underground. Such ulcers are often painless, but nevertheless, such a sight seemed positively medieval. I suspect his employers were totally unaware, and possibly even his workmates, if he kept his vest on.

Another illustration showed an erect index finger with an ulcer on the tip. Macfarlane explained it belonged to a dentist in Edinburgh who was a personal friend and who telephoned him at three o'clock one morning in a panic, having finally realised it was a primary sore of syphilis, which had resulted from contact with a patient's tonsil in the days before dentists wore gloves. The donor must have caught it indulging in some form of sexual practice which has a Latin name.

Fortunately, such medical oddities are now almost extinct, certainly in the western world, although syphilis is said to be making a comeback. Of course, it was incurable until the discovery of penicillin, but after the primary sore and then the fleeting second stage, with its rash and vague ill health, it could remain dormant for up to thirty years until the final stage. The standard treatment in the old days was with mercury salts, themselves toxic, and their apparent efficacy probably relied upon the often delayed latent period. Congenital syphilis could lead to horrific deformities in innocent children. I have a photograph of a teenage boy in an Edwardian suit with four holes in his face where his eyes and nose should be and exposed teeth but no lips.

***

Ward twenty-one, the Dermatology ward, was in the basement of the hospital near the pharmacy and CSSD (Central Sterile Services Department), and thus not exposed to natural light. Whether this was by accident or design I do not know, but some skin conditions are light-sensitive. 'Skins' remain a mystery to most doctors, which is a pity because in general practice, consultations for skin complaints comprise a significant number. We used to have a saying—'A spot is a spot is a spot.' As for treatment, if it didn't respond to coal tar, try steroids.

***

Inpatient psychiatry was at the St Nicholas Hospital, but the RVI had a Department of Psychological Medicine where outpatient clinics were held. The professor was Martin Roth, later knighted, who was appointed in 1956 and who stimulated the growth and development of a department with an international reputation. Roth was born in Budapest, the son of a cantor in a synagogue, and came with his family to the East End of London at the age of five. What hair he possessed was black, and with his bald pate and distinguished nose, he looked the intellectual he undoubtedly was. He was precise in both dress and speech. He thought before he spoke, but then the sentence was perfectly constructed. He became the first president of the newly formed Royal College of Psychiatry in 1971 and left Newcastle in 1977, tempted by the first Chair in Psychiatry at Cambridge.

Sitting in with him on an outpatient consultation was an education in itself, for the discussion afterwards in the absence of the patient would cover a wide range of subjects, rather as I imagine an Oxbridge tutorial to be. At one clinical demonstration, he presented a patient with what we then called 'manic-depressive psychosis', now more gently labelled 'bipolar disorder'. After relating the history, they brought the patient to us from an adjoining waiting room. The man had been a porter in the House of Commons and an MP had taken pity on him and handed down an unwanted tuxedo. When the poor man entered he was wearing this now somewhat soiled suit, which he used as normal apparel, as he possessed no other. It was an amusing sight to us, but naturally we displayed no outward signs of such in his presence. I remember the prof subsequently commenting that the suit was probably appropriate for the Commons, because during his manic phases he did have delusions of grandeur.

When I was a senior registrar, I had a bizarre experience with Sir Martin which made sense when I read Lord Walton's autobiography, in which he states that in Newcastle he was known as 'the late Professor Roth'. I presume that after 1971, when he was the president of the Royal College of Psychiatrists, he often travelled to London. The Newcastle taxi drivers got to know him and learned to call well before the appointed time because he would invariably forget or misplace his passport, tickets or overcoat. This behaviour tallies with my story. As SR, it was delegated to me to arrange the final examinations held on an emptied ward. Although the patients had surgical diagnoses, the examiners were a mixed bunch.

This was an intentional move, recently introduced to achieve 'fairness'—that is, to prevent a super-specialist from demanding more than average knowledge about a particular hobbyhorse. Because examiners would rotate to several wards throughout the day, punctuality was crucial. I was showing the various patients to the examiners when up pops Sir Martin, only a little late. Very politely, he asked if I would perform a favour. Would I please walk about two hundred yards down the corridor, seek out his Mercedes in the front car park, close the windows, switch off the engine and return the keys to him?

'Oh, and I almost forgot, there is a Grundig tape recorder in the back seat, still running,' he added. 'Could you please switch it off and retrieve it as well?'

It was all said with a lovely apologetic smile. I suppose this degree of eccentricity correlates with one's IQ. He died in 2006, aged eighty-eight. A lovely man.

The other main teacher in psychiatry was Bobbie Orton, a diminutive, busy, fast-talking man. Students view most psychiatrists as being a bit odd, to say the least, but he was a down-to-earth, sensible person who was a general physician until after the war, when he entered the specialty because of the demand. There was an insufficient number to deal with the psychological consequences of warfare in combatants and hence he 'read 50,000 pages of psychiatry as rapidly as possible and got on with it'.

Chapter Twelve

The Final Cut

At the beginning of final year in 1965, Roger and I took a room in 'Wor Hoos', Geordie for 'Our House', where Frank Bone and another chap in our year, Norman Altham, had lived for some time. There were ten or so rooms, a communal common room and TV and a kitchen. There was no rift with George and Jim, but we felt that in this, our final year, with two sets of crucial exams, so-called 'housekeeping' would be simpler than in Glenthorn Road.

Norman had joined our year from the one above after failing his exams. He was a lovely chap, but unfortunately, his life was blighted by mental illness. Most of the time he appeared normal, if a little hyperactive. I certainly never saw him depressed, but on occasions he would become hypomanic. This took the form of playing either the trombone or piano at full tilt for what seemed like ages. He actually played trombone in a jazz band, but it was the piano playing that was a sight to behold. He would belt hell out of the upright in the house at a furious pace, with one foot tapping like a fiddler's elbow; nor was it rubbish, it was good trad jazz. I assumed he suffered from the aforementioned bipolar disorder, but he related an experience which gave me the creeps. He had been going through a religious revival and became friendly with Jack Bennett, the vicar of St Thomas, across the road from the university, and chaplain to the student fraternity. One Sunday evening after church, he was talking to Jack on a religious topic in the latter's room. On bidding the clergyman goodnight, as he opened the door, a voice said, 'Norman'. On turning round, instead of Jack sitting in his chair, there was Jesus Christ himself, dressed in robes and with blood dripping down from a crown of thorns. He described this in a balanced, matter-of-fact way. I didn't know how to respond because, to me, visual hallucinations of this nature were diagnostic of schizophrenia and had a poor prognosis. So I still do not know his diagnosis and it no longer matters, as Norman died a long time ago.

In December 1965 we took the first part of finals, in obstetrics and gynaecology. There were written papers, clinicals and vivas in both subjects. Like other exams, there were rumours, possibly mythical, about how to fail! It was said that if you got Professor Russell (which I did) in the gynae section, when one would certainly be asked to perform a vaginal examination on some poor woman who had supposedly volunteered (!), then one had to remove the examining glove slowly in a downward direction and away from him. Otherwise, if there was the slightest risk of lubricating jelly flying in his vicinity as the glove snapped off, you were doomed. All rubbish designed to scare us, passed down from year to year. Or, knowing Russell, possibly not.

For my viva I got Derek Tacchi and Harvey Evers. I have already mentioned the former and his successful private practice. The latter I had heard of by reputation but never encountered. Harvey Evers was the professor before Jake, but long before he retired in 1958 he was a Newcastle institution in the field of obstetrics, owning a private hospital in Jesmond referred to by staff and students as 'The Golden Gates'. He had a reputation for clinical and teaching brilliance, as well as sartorial elegance and a taste for fine cars, such as a 20/25 Rolls-Royce coupe and a Bentley Continental. My landlady in Brighton Grove, Mrs Stuart, told me she knew of ordinary women who saved up to have their baby delivered by him privately.

Derek Tacchi introduced me to him. Naturally, I was apprehensive, but he gave me a lovely smile and put me at ease by simply asking me to describe the three stages of labour. I could hardly believe my luck and the rest was plain sailing. Traditionally, successful candidates made a round of the consultants' houses, the professor excluded. We went briefly to Tacchi's and ended the night at Miss Kerslake's, singing songs to the accompaniment of her aged mother at the piano.

The year turned and as 1966 began, we started the run-in to finals in June. Statistically, we knew the odds were in our favour because most of the failures had either left of their own volition or been filtered out in the early years. But there were always a few who tripped up and would have to resit in December.

We had now completed the syllabus and it was a question of polishing our skills at hospital practice and revision of mainly medicine and surgery. There was a separate exam in ophthalmology, but it didn't count for much. The long medical case occupying three hours could be unpredictable—although usually an adult case with pathology, it could be a child or a psychiatric case, as could any of the questions in the written papers, so nothing could be ignored and left to chance.

Between January and June, everybody had one last month in residence in a peripheral hospital, doing medicine. John Davison (later my best man) and I were sent to Darlington Memorial under the supervision of Dr Edmunds and a newly appointed physician, Joe Hampson. We nicknamed the two elderly sisters who manned the female ward Elsie and Doris (after the radio comediennes, surnamed Waters). Their 'office' was an oval contraption in the centre of the ward, encased in either glass or Perspex, the like of which I never saw before nor since.

Despite working in a district general hospital in a small provincial town, the senior surgeon Mr McKeown had a national reputation as a master technician, especially for his so-called three-stage oesophagectomy for cancer of the gullet. Belfast-born and trained, he had been first assistant to Prof Ian Aird at the Hammersmith before coming to Darlington and Northallerton. However, we had to adhere to medicine, and so could not benefit from observing his surgical prowess.

There was a nice quiet library in which John and I could revise. Our main textbooks were Davidson's *Principles and Practice of Medicine* and Bailey and Love's *Short Practice of Surgery*, the word 'short' being a misnomer. They were supplemented by several others, two of which, on haematology and cardiology, were by RVI consultants. Bailey and Love, first published in 1932 and updated editions of which are still in print, was nicknamed 'Steatorrhoea' by generations of medical students, this word being used for the pale, bulky and very offensive stools passed by patients who cannot absorb fat. The cover of this 1,200-page book was a buff colour, hence pale and bulky, and the contents were deemed offensive!

Finals occupied the last three weeks of June, traditionally ending on the Friday of the week of the annual Town Moor Fair, the Hoppings. Several prize exams, the result of bequests by long-forgotten alumni, preceded them. Entry for these was not obligatory, but most of us 'had a go' to gain practice at essay questions and clinicals. It was obvious, however, that the examiners regarded them as a chore to be dispensed with as quickly and dismissively as possible, which was disappointing.

There were three-hour written papers in both medicine and surgery, and for the first time a multiple-choice paper of the same duration. This was novel and used to compare our results with those of the year below when the time came because they would be the first to complete the new curriculum. I presume the hope was that the results would be either statistically insignificantly different or to the advantage of our successors, but we were never told.

In the Medicine long case, one had to write the full history of a patient and the findings on physical examination. You then created a differential diagnosis and suggested appropriate investigations, some of which were then released, in order to narrow down the presumptive diagnosis. My patient was a seventeen-year-old youth and the only abnormality I could find was that he was clinically anaemic. This concerned me at the time because I felt that I must have missed something, but in retrospect, anaemia in a lad of that age is uncommon, with a multitude of possible causes, and I reassured myself that the examiners were looking for a logical differential diagnosis and the correct investigations.

The surgical long case comprised twenty to twenty-five minutes with a patient, before being taken aside by the examiners. The discrepancy in time between the two major specialties was because in the medical case, one was expected to cover every system in the body, including a full neurological examination, whereas in surgery one concentrated on the relevant system. My patient had a cancer of the caecum, that portion of the large bowel low down in the right side, from which the appendix dangles, and it was easy to feel. I was examined by Peter Dickinson, on whose ward I had been a student, and Angus Hedley-Whyte, a retired surgeon. Like Harvey Evers, he seemed delighted to have been asked, and was benign. I expected to be asked to recite the history but he took me well away from the patient and simply said, 'What do you think he's got?'

'Carcinoma of the caecum, sir.'

'I think you're right—come this way,' he said, and took me to a series of short cases.

He then asked me a few simple questions of applied surgical anatomy and beamed that he liked to see a young man who knew his anatomy. I sailed through the next few 'spot' diagnoses, all the while Peter D smiling reassuringly. I remain very grateful to Mr Hedley-Whyte, whom I never again encountered. His most striking feature was a prominent, bulbous nose. Ian McNeill, one of my chiefs, said that when they were giving out the noses, Hedley-Whyte thought they said roses and asked for a big red one! When I worked for Brian Fleming, he came in chuckling one day, having met the old man, who was now retired to Alwinton in Northumberland. Hedley-Whyte informed Brian that he was visiting London for a few days and asked would he be so kind as to look after the family silver for him.

The ophthalmology exam was a farce. The chief, Vernon Ingram, had never taught us, but he examined us. The man who had done virtually all the teaching, the tall, bearded James Howat, told us to use the ophthalmoscope by gradually approaching the patient's eye and flicking through the different lenses to examine the lens, the contents of the globe and finally, the retina. Vernon Ingram put his hand on each candidate's shoulder and instructed us to examine the retina. He then pushed us down, and no sooner had we focused on our object than he pulled us back using the hand still on our shoulder and demanded to know what we saw. I would love to know how many replied 'Nothing, sir.' But that was it.

And then Neil Manson, an ex-demonstrator and now SR in Eyes, played a dirty trick on the morning that I took the exam. He asked me to examine an old lady's eye. She had a fixed irregular dilated pupil with clouding. Although I had never seen one before, it looked just like an acute glaucoma. But this diagnosis was a surgical emergency, not to be delayed or blindness would result. Surely it was too much of a coincidence for her to have just appeared? So instead of sticking to my guns, I blurted out some alternative diagnosis. Manson sneered and told me it was acute glaucoma, an emergency, to never forget the appearance and to get out now! It was pointless to splutter that I thought this all along and wondered why you were not getting on with treating it, instead of delaying for the sake of a pathetic little exam that didn't really count for much. And who wanted to do eyes, anyway? Well, Johnnie Howe, actually, in our year—and he made a fortune out of it! Incidentally, the aforementioned James Howat worked his socks off to reduce the waiting list for cataract surgery by doing extra lists on evenings and weekends, *gratis*, only for the list to rise again during one of the several industrial disputes by ancillary staff.

In the surgical viva I got Frank Walker, shortly before he left in the fit of pique previously described when he failed to obtain the Chair in Surgery after Prof Lowdon's untimely death. Frank was obviously bored. There was a merchant seamen's strike at the time and in keeping with his at times weird sense of humour, Frank asked what I thought could be the surgical consequences of such a strike. This was obviously an opportunity for me to dictate the course of the viva. If I had kept calm and chosen any subject I felt knowledgeable about, I could have manipulated it into the situation. But I was thrown and somewhat bewildered, naturally, so I hesitantly said that if there was a violent confrontation with the police, there may be head injuries. Fortunately, having dropped myself in, I knew enough to extricate myself.

At long last, after five years of study, the whole year congregated in the foyer of the Medical School at 5pm on the Friday to hear the registrar read out the results. He started by saying that as a result of the final examination for the degrees of Bachelor of Medicine and Bachelor of Surgery, the following have satisfied the examiners; with First Class Honours and a Distinction in Medicine and Surgery—Niall Edward Foster Cartlidge. For the moment, anxiety about one's own fate abated as the cheers and applause rang out and it seemed to take at least a minute to die down. There hadn't been a First in Medicine since Tophie Wynn, and he was now middle-aged. Then, he announced, with Second Class Honours... and together with five others, there was my name. I was too numb to hear the remainder, but poor old Forbes, Ralph and a few others had failed (they all passed resits in December).

When the excitement died down, we were directed into the Howden Room, where the examiners and the new Dean, Henry Miller, congratulated us and was positively beaming, pumping our hands as if we were his own progeny. Our undergraduate careers had started and ended in that same room.

I telephoned Mary and we arranged to meet for the usual celebrations, which by long-established custom, commenced in the Brandling pub. By previous arrangement, my mother remained in the shop after it closed, awaiting the telephone call. She was thrilled when I explained that Second Class Honours was actually a good sign, and not one of having scraped through, with only six of us having achieved it.

The next step was the house job, the compulsory year of medicine and surgery in a hospital. It was officially called a pre-registration post because only after 'satisfactory' completion of such a term could one's name be included on the Medical Register and one would become a legally entitled medical practitioner. Such posts were guaranteed because their number nationally matched the number of graduates. By custom, medical schools supplied their own region, although in theory, one could apply to anywhere in the country. Those with ambitions to be consultant physicians or surgeons applied for one of the twenty-four posts in the RVI, a springboard to the first rung of the training ladder. I had already decided on a career in surgery, if possible, and only applied there. This might seem like overconfidence, but the Catch-22 situation was that the RVI always announced the successful applicants after finals results. Most of my fellow students had already been promised a job elsewhere, conditional on passing the exams. If any of us provisionally accepted such a post, we would have been disqualified from the RVI.

Of course, it was still early to make a final career choice, and competition up the slippery slope was fierce. But surgery attracted me because I had particularly enjoyed anatomy and pathology and my experience of the surgical firms was that we students were embraced as temporary members of a happy family. The teaching was excellent and taken seriously and the long-established but sadly now obsolete 'firm' system described the camaraderie and teamwork perfectly.

To my relief, I was appointed and in about three weeks would join Messrs Dickinson and McNeill to do six months' surgery. Only this first job was guaranteed—one had to reapply for the second six months, but in practice, unless one's performance had been poor, it was routine to be reappointed. Next was the graduation ceremony to receive our degrees. As you may recall, when I started in 1961, King's College, Newcastle, was part of the University of Durham. But in 1963, Newcastle separated to become a university in its own right, so they gave us the choice of either a Durham or Newcastle degree. Most settled for Durham; I suspect due to a combination of snobbery, nostalgia and a more attractive environment for the ceremony. It was held in the castle, next to the cathedral, on the summit of a peninsula surrounded on three sides by the River Wear, forming what Pevsner described as possibly the best architectural view in Europe, and he was no biased Geordie (to be pedantic, Durham is not strictly Geordieland, but the accent is very similar).

Mary and I made our way to Durham by bus, and there we met my mother and Aunty Elsie. These two war widows, separated in age by eighteen years as the eldest and youngest of three sisters, often took holidays together now their offspring had left home, and Aunty Elsie always seemed as proud of my success as if I was her own son.

We wore ermine-edged hoods of palatine purple to our gowns, hired from Gray's of Durham (still extant). Only members of the academic procession wore mortar boards. We each shook hands with the Chancellor, the Earl of Scarborough, as they announced our name. Only later did we learn that Norman Altham, our fellow occupant of Wor Hoos, owed his presence to the kindness of a couple of fellow residents who spotted him catching a bus and alighting at the next stop, looking agitated. They accompanied him all the way to Durham and ushered him into the castle, where members of our year took over and pushed him forward, sallow and sweating, to receive his degree.

At last we made our separate ways home. In all the excitement, I doubt if it struck us at the time, but for some of us, it was a final farewell. Indeed, I have not seen at least eleven of our year since. One of them, Mukunda dev Mukherjee, who was a colleague on the same dissecting table for eighteen months, and who spent his entire undergraduate career as a resident of International Methodist House, bade me farewell on Elvet Bridge in Durham, and I can see him now. He hailed from Calcutta and from previous conversations, I fully expected him to return there and spend the rest of his life ministering to the poor. But he had done his elective in the States, and Satan had tapped him on the shoulder. As my friend George Bone said later, bitterly, 'There was I, struggling to get the necessary requirements to practice in the States, when Mukherjee, who I expected would be sipping goats' milk on the banks of the Ganges, is driving a white Cadillac with his American wife as a GP on the outskirts of Boston.' You can't blame him. In fact, he already had an elder brother in the US and so probably made up his mind a lot earlier but kept his counsel. Much later, I was to learn of the almost unbelievable denouement of Mukunda's career.

Chapter Thirteen

Doctor in the House

T he fateful day came all too soon, when twenty-four of us gathered together in the RVI boardroom to receive our bleeps, the keys to our rooms and a handbook and then disperse to our various posts. The hospital staff dreaded the brief hiatus between the outgoing and incoming 'house'. Traditionally there was a raucous party on the last night of the 'old' house but, of course, somebody had to be on call. The domestic staff cleaned up and changed the bed linen and the nurses waited with trepidation for the new boys (and girls, of whom there were three).

Bill Cookson, Arne Walløe and I made our way to Pavilion One, the base of the Dickinson/McNeill firm. It was at the Leazes End of the hospital, hence the operating theatres along the corridor were the Leazes Theatres, which years before had looked out onto fields on which cattle sometimes grazed. They were also the site of a story about the great neurologist Henry Miller when he was a student. Apparently, he was well known for having little manual dexterity. One summer's day, a patient was anaesthetised for an examination of the bladder. Previously, the man had received a suprapubic cystostomy, presumably for prostatic obstruction. In this procedure, the bladder is drained directly via a catheter placed surgically in the lower abdomen rather than the penis. The surgeon inserted the cystoscope into the bladder, this time via the penis. The bladder was then filled with fluid to distend it and obtain a view, and Henry was asked to look through the instrument and describe what he saw. There were, however, two added features.

Firstly, because it was a hot day, the normally frosted windows of the theatre had
been raised. Secondly, on wiggling the instrument about, Henry inadvertently
pushed its tip through the small, barely healed hole in the abdominal wall, so
that, unknown to him, it was peeping outside the patient's body. Furthermore,
the instrument was an old-fashioned Brown-Buerger, the lens system of which
created an inverted image and had to be interpreted as such. So when asked what
he could see, Henry replied with the immortal words, 'A cow, upside down, in a
field.'

When we were greeted on Pav One, we found to our dismay that two of the six
months were to be spent on the orthopaedic wards, and I drew the short straw.
Off I trotted to meet the delightful Sister Shaw of W18, the female ward, where
she had been acting as house surgeon for the last week, and hence was particularly
pleased to see me. Unlike some of the other ageing nursing spinsters, she had a
sweet nature and was kind and helpful.

Most of the female beds were occupied by elderly women with fractures of
the femoral neck, usually due to falls associated with osteoporotic bones, and the
male beds by the victims of road accidents with fractured long bones of the leg.

For some stupid reason I looked forward to the first occasion on which I would
be up all night with an emergency. It was not long in coming. A nineteen-year-old
student from Thailand hitched a lift on the Great North Road near the Town
Moor and within the first quarter of a mile the car was in a collision, resulting
in our student receiving fractures of both femurs. After resuscitation, he went to
theatre to have the bones fixed. In those days, all blood for transfusion had to be
fetched from the bank by a doctor, whatever the time of day or night. I obtained
a couple of bottles (no plastic bags then!) and started to erect one, only for the
anaesthetist to enquire, quite correctly, if I had checked the hospital number.

'No,' I replied.

'Well, you should always do it.'

'I know,' I said—and then, cheekily, 'but I don't expect there's more than one
Viraf Nimmanvathana in the hospital tonight.'

He was not amused.

To digress a little, on that first day when we dispersed to our stations, John Coles, a fellow houseman whom I knew well because we were on the same dissecting table, went down to the Accident Room to be greeted by a staff nurse, who said she was pleased to see him, 'because there is a BID in Room One.'

'What's a BID?' asked John.

'Brought in dead,' was the reply and indeed, in the room was the corpse of a British Railways employee who had been struck by a train.

It sounds like a sick joke, but on the epaulettes of his uniform were the words, 'Look Out'.

On the afternoon of the 1966 World Cup final I was catching up with work on the ward. TV sets were only present in the day rooms in those days, not elevated on the walls as they have been since Nightingale wards were abolished. All I knew was that England were losing 1-0. Although as a student I occasionally watched Newcastle United with friends, I was no longer particularly interested, but I eventually wended my way back to the residence to find the common room packed and my latent patriotism reignited. Even dear old Martha le Dune, a registrar then, bespectacled, with unfashionable clothes and hairstyle and an awkward schoolmarmy gait, punched her arm in the air in triumph at the final goal.

Some of the old ladies with fractured necks of femur could theoretically have been treated non-surgically by a form of traction, but healing took about three months, during which time there would have been significant mortality and morbidity due to the complications of prolonged bed rest, such as deep vein thrombosis, bed sores and pneumonia. The standard treatment was surgical, either in the form of a pin and plate to fix the fracture or removal of the head of the femur and replacement with a metal prosthesis, like the modern hip replacement, but less sophisticated. The aim was to achieve early mobilisation and prevent the complications mentioned, but in practice, it was difficult. Many of these patients were not very mobile beforehand and with only limited physiotherapy facilities, they often remained as 'bed-blockers'. Although there were more convalescent beds than there are now, usually in small cottage hospitals, they still needed to have some degree of mobility before discharge, and this took time.

Every week the head porter came to the ward with the 'two-month book', into which Sister entered the details of any patient hospitalised for this length of time. It was then countersigned in what seemed a serious, formal ritual. It was years before I realised that after two months OAPs get their state pension either stopped or reduced until they are discharged, a custom which is still in force, I believe.

One day, an old woman recovering from surgery to her fractured femur suddenly collapsed and died from a pulmonary embolism. Trauma and surgery spark off chemical processes in the body that make the blood clot more readily, especially when combined with inactivity, and thrombosis in the deep veins of the leg or pelvis was a common postoperative complication. Such a clot can break off and flow to the lungs and block the blood supply. If it's large enough this can cause sudden death or, if smaller, pleurisy and the coughing of blood. Since the 1970s, almost all patients undergoing surgery receive low doses of anticoagulants to try to prevent thrombosis, unless there are contraindications, and fatal pulmonary embolism is much less common. In the 1960s, however, it was seen more often, and this case was no different—except that it occurred just before one of the rare visits of the senior consultant, Mr JK 'Jake' Stanger. Sister Shaw was pleased to see him, as they were friends of some thirty years, and soon got down to reminiscing. Jake included me in the conversation and, after bemoaning the 'tragedy we have just had', told me a story I could not believe I was hearing, especially as it was related in such a matter-of-fact manner, with Sister giving the occasional nod and benign smile as he looked from her to me in the telling. Apparently, sometime in the early fifties, a female houseman was on the ward. Over a period of time, a cluster of elderly women with femoral fractures died suddenly. I was not told whether they were subjected to autopsy or referred to the coroner, except that their deaths were unexplained, apart from the fact that they were old. Sister must have looked into their details further and discovered that every single one lived alone and had no close relatives or visitors—on reflection, a remarkable piece of detective work. One day, the houseman was interrupted in the process of giving an old lady an intravenous dose of morphine. It seems that the situation was dealt with totally internally. The reason given was that these interventions were acts of mercy to old women who had no future. The solution was that it had to stop—and no further action would be taken! There was a pause in the conversation, presumably for it to sink in. Then Jake smiled at Sister.

'She emigrated to Canada, you know,' he said. 'I still get a Christmas card from her every year. She was one of the best housemen we ever had!'

I have told one or two people this story. I only commit it to the page because both Jake and Sister are long dead as, almost certainly, is the culprit, who in any case would be untraceable. There are shades of Harold Shipman, but presumably a different motive. Coincidentally, when Shipman was arrested, an anaesthetist colleague of mine said he had been a fellow student of the serial killer at Leeds and was on the same dissecting table. The following day, my colleague's parents were besieged by journalists seeking an interview with their son. Neither he nor his parents ever found out how the newspapermen identified the connection so rapidly and accurately. Incidentally, there was a belief by some nurses that W18 was haunted. I suspect most large hospitals have similar superstitions. It was said that many years before, a nurse inadvertently gave a patient an overdose of morphine which proved fatal, and was so filled with remorse that she committed suicide. The rumour was that whenever a nurse drew up a syringe of morphine in the treatment room, she received a tap on the shoulder from an invisible hand to make her check the dose.

Our salary that year was £700, but our accommodation and catering were free. The 'house', or doctors' residence, was next to but separated from the nurses' home, and was approached from the main hospital building via a lovely glass conservatory. It was always refreshing and cheering to walk through it each morning and early though it was, the plants had usually just been watered. The Edwardians certainly had style and weren't frugal in their planning. Some psychopath in the year below us, while drunk, threw a stretcher pole from the second floor above, right through the roof of this structure. Fortunately, it was at night, with no-one passing, otherwise they would have been killed. The culprit became a professor of orthopaedic surgery of national repute.

We each had our own room with a bed, wardrobe, bedside table, desk and chair. There were no facilities for married couples. We had our own washbasin, but bathrooms and toilets were at both ends of each floor.

Lunch was served from 12.30pm to 2pm, and with the exception of theatre days, we could usually manage to get to the dining room and enjoy a cooked meal with waitress service. The ladies, nearly all middle-aged, wore yellow dresses and caps and white aprons. Most were long-term employees who were treated like friends, referred to by their first names and proud of their association with the RVI. The ward nurses endeavoured not to call us away unless it was necessary. On reflection, it seems like a different era—very few working people now sit down to a cooked lunch and the idea of having it served sounds positively luxurious, but we had an onerous rota, both housemen being on duty on reception about one day in four, as well as sharing the remaining nights of the week. A weekend off began at about 2pm on Saturday. On one occasion, both my housemen colleagues took their holidays consecutively, and I remained in the hospital for a month without leaving it. A colleague in the year above, Hugo Rowbotham, whose parents were a local neurosurgeon and his child psychiatrist wife, served as an SHO in ENT while we were housemen. Even under normal circumstances Hugo sported a bouffant hairstyle, but it had grown out of control and was beginning to resemble a tiered wedding cake. One of his consultants remonstrated with him, saying it was time he got a haircut. Hugo pointed out that it was six weeks since he had left the hospital and that if they could give him the time, he would be delighted to oblige.

One weekend, a man in his thirties was admitted with an 'injured leg'. He was on a rock climb and jammed the leg between some rocks before being released by the mountain rescue team. No fracture was detected on x-ray, but his pain and distress persisted. It eventually became clear that his pain was psychological in nature, but treatment was difficult. He was single and of low self-esteem and frequented the favoured pubs of the rock climbing fraternity. He had bought all the gear and had read the books, but when at last his ability was tested, he panicked and feigned injury. All this was revealed when a psychiatrist gave him intravenous thiopentone, which was usually employed as an anaesthetic but when given in small doses was known as a 'truth drug'. It certainly worked with him, and very impressive it was. A pit deputy from Washington was admitted after a fall of stone and had been permanently paralysed below the waist. Coincidentally, a miner from the same pit was already in the ward with a knee injury. His reaction when he heard about his senior colleague was that the man was a bastard and deserved it. So much for the legendary comradeship of the miners.

One day, an old lady called me over and sternly and convincingly accused me of projecting pornographic films onto the ceiling the previous night. Of course, she was delirious and hallucinating, but she was so cross that I actually felt guilty as charged, and felt myself blushing!

It was lonely being the sole houseman on orthopaedics, and the unit seemed leaderless, like a ship without a rudder. It was with relief that after my obligatory two months, I transferred to Pav One as a houseman to the Dickinson/McNeill firm. Both men had a huge influence on me and my career.

Peter D, as he was known, was the senior by a few years, a tall man with a measured, dignified walk and an extrovert who was a steady, careful surgeon with a great deal of common sense. He was in the process of building what became a flourishing private practice and in these early years took his holidays on the Northumbrian coast so that he was always readily available. He always drove either a Jaguar or a Mercedes. Ian McNeill (Mac) was slightly shorter but well-built, with broad, handsome features and wavy grey hair. He was more introverted, classed as a surgical intellectual—if that's not an oxymoron; extremely widely read and technically skilled, with the gift of significant manual dexterity. As befitted his quieter nature, he drove a Rover.

Both were local graduates and were easygoing. They got on well, took teaching seriously and gained both the respect and the best out of their staff by example, without needing to impose strict discipline.

The working day for the housemen started with bloodletting. Every patient receiving intravenous fluids—that is on a 'drip', usually post-operatively—had their electrolyte levels measured daily and other blood tests carried out as required. There was an early 'business round' by the registrar or senior registrar, usually the one who was not in theatre that day, when routine management decisions were taken. Both consultants had subspecialty interests in peripheral vascular surgery, which was still in its relative infancy in the 1960s. For practical purposes this meant two major operations. One was the repair of abdominal aortic aneurysms—that is, the replacement of the abnormal ballooning of the main artery in the abdomen, the aorta, which usually occurs between the arteries to the kidneys and the division into the iliac arteries supplying the pelvic organs and then the legs, by an artificial graft made of Dacron polyester fibre. The second operation was a bypass of a blocked or narrowed aorta or iliac arteries by a similar graft. Aneurysms are more common in men, smokers and those with high blood pressure, and are usually not symptomatic unless they rupture, when they are fatal either suddenly or within a few hours. In those days they were discovered either by chance, or when patient or doctor felt a prominent pulse in the abdomen or when calcification of the aortic wall was found on an x-ray taken for another reason. In only a few patients was it revealed because of symptoms such as backache. Nowadays, ultrasound scans often uncover them, again usually when performed for another condition, such as gallstones. But now that preventive surgery is safer—and can often be done using a less invasive procedure, the insertion of a specially designed stent—the government has agreed to a national screening programme for men aged sixty-five and over. About 4,000 elderly men were dying every year from ruptured aortic aneurysms. Patients can survive a rupture, but the odds are stacked against them—about half die in transit to hospital, and of the rest, the average survival rate is less than a third.

Patients with narrowed or blocked iliacs or even aortas present with muscle pain on walking, starting in the calves and working up to the thighs and buttocks. This eventually causes them to stop at varying distances and rest until the pain abates and they can set off again. This symptom is called intermittent claudication (from the Latin, *claudicare*, to limp), and the narrowing is caused by atherosclerosis, so once again the risk factors are smoking, hypertension, diabetes and genetic predisposition. You might wonder how anyone can walk at all with a blocked artery as opposed to a narrow one, but the process takes time and small arteries above the diseased section gradually open up and create a 'collateral' circulation, in the same way that if a river is dammed, small streams above the dam can compensate and take the flow. Patients with this high distribution of disease were about a third of our clientele, the remaining two-thirds being generally older and with a blockage or narrowing in the artery to the thigh in a muscular canal between the groin and the knee. Once again, the presenting symptom was intermittent claudication, affecting the calf muscles, but in this latter group there was the paradox that whereas spontaneous improvement with time was more likely, as the collateral circulation took over, they were more likely to lose the limb than the former group. Unfortunately, the results of reconstructive surgery for these patients were poor and on our unit it was rarely performed in those days.

Although the first group were less likely to lose a limb—unless, of course, they subsequently developed the lower disease—their symptoms were much less likely to improve spontaneously. As they tended to be younger and still capable of work, they were offered surgery, with its prospect of good long-term results, at the price of a major procedure with a small, but significant, risk to life. Both operations took at least three hours.

***

On Monday afternoon there was a 'grand round' of all patients with the whole team, including attached medical students, and afterwards in the seminar room we planned the elective work for the following week. We were assisted by the 'bed board', a suspended wooden board showing the beds of both wards. Each bed had three coat hooks and on these were coloured discs, red in the centre if occupied and green if empty. Discharges were predicted and a disc with the appropriate day of the week attached at the foot of the bed and a planned admission attached at the head.

Usually, three beds for each sex were allotted for reception days. The planning was done with nursing staff and secretaries present, and the latter produced the waiting lists so that suitable cases could be sent for. The time taken to perform the operations depended on the procedure, and this had to be taken into consideration so an operating list could be completed in the time available.

Reception days were a fixed weekday each week, and one of the weekend days on a rolling rota. There was a full ward round each Saturday, followed by case presentations with discussion in a side room.

Because attempts at vascular reconstruction below the knee were rarely attempted due to the high failure rate, amputations were common, usually above the knee. Rehabilitation is much more difficult with such amputations compared to the below-knee type, especially in an elderly population, and this was not helped by heavy, cumbersome prostheses, the design of which did not seem to have progressed since both world wars. The limb-fitter was usually a 'failed' surgeon who didn't meet the patient until discharge, an unsatisfactory situation. It really was a Cinderella subject. Fortunately, great strides have been made in vascular surgery since. But it has taken two major wars, in Vietnam and the more recent conflict in Afghanistan, in which young men have been rendered limbless, to drive the development of light, practical prostheses.

No pharmaceutical preparation could help peripheral blood flow to a significant extent, so our in-patients were prescribed that well-known vasodilator, alcohol, on the NHS. They had a choice of whisky or brandy, but it was stored in non-proprietary labelled bottles, and I doubt it was of good quality.

Although most surgeons labelled with the prefix 'general' offered a subspecialty interest within the broad field of surgery, certainly in the teaching hospitals, the bulk of the work encompassed a wide variety of conditions and operations. Today, there are breast and endocrine, upper gastrointestinal, colorectal, vascular and urology surgeons. In the 1960s, all these were within the province of the general surgeon. In addition, most soft tissue trauma, including that of the hand, was dealt with.

Peptic ulcers were common and long before the Helicobacter pylori organism was implicated in causation, severe symptoms or complications such as bleeding were treated surgically on the grounds of 'no acid, no ulcer', and secretion of gastric acid was suppressed either by division of the nerves stimulating it (vagotomy), or by removal of a large portion of the stomach itself (gastrectomy). The standard operation for breast cancer was radical mastectomy—a mutilating procedure involving not just removal of the breast, but also the underlying muscles, together with clearance of the lymph glands in the armpit (axilla). This was on the cusp of being replaced by simple mastectomy (that is, the breast alone) followed by radiotherapy, and over succeeding years, patients have been managed with lesser surgical procedures.

The commonest emergencies were encompassed within the term 'acute abdomen'. Examples are appendicitis, perforated duodenal ulcer, gallstone colic and inflammation of the gall-bladder (cholecystitis), small and large bowel obstruction and pancreatitis. Abscesses were frequent, often affecting the rear end—ischiorectal, close to the anus itself, and pilonidal, just above the cleft of the buttocks, otherwise known as 'jeep disease' due to its predilection for hairy, sweaty men bouncing up and down in such a vehicle, presumably originally GIs in the Pacific. It is actually caused by hairs being driven into the area and the term 'pilonidal' means a 'nest of hairs'. I have seen it in pretty blonde women, but it's more common in dark, hairy men.

Appendicitis is essentially a clinical diagnosis—in other words, there is no specific test for it. The history is important and usually follows a clear pattern, but in the final analysis, the crucial finding is of localised tenderness in the right iliac fossa (RIF), low down in the abdomen and specifically at McBurney's point, two-thirds of the way along a line between the umbilicus and the iliac crest bone. But there are many exceptions. I once saw a woman who was washing up and suddenly felt unwell, fell to the floor and started shaking. This lasted ten minutes before she got up, but had a pain in the RIF. Despite this weird story she had a rip-roaring inflamed appendix at operation. If there was doubt about the diagnosis, and especially in females of child-bearing age in whom there are several other possibilities, a period of observation was permissible, but to allow an appendix to proceed to perforation on the ward was regarded as disgraceful. Once an appendix perforates, the morbidity increases significantly, as indeed does the mortality, although that would be rare today. The generation before mine had qualified before antibiotics and had seen previously fit young people (and in general, appendicitis is a disease of the young) die. Consequently, there was a tendency to operate early rather than later, with the result that a significant minority of removed appendices were lily white (that is, uninflamed) especially in females, because ovulation pain can sometimes resemble the condition. Most patients seemed to accept this with equanimity, but no-one desires unnecessary surgery and over the years attempts have been made to fine-tune the diagnosis without resorting to the knife. In the USA, where CT scanning is more widely available, this is now used as the deciding factor, and it is creeping in here.

Furthermore, most appendices are removed these days by keyhole surgery (laparoscopically), which entails an even smaller incision than the now 'old-fashioned' one. I was told recently that a specialist registrar called out his consultant in the night because he couldn't complete the procedure by the keyhole method. When advised to convert it to the open technique, he admitted he did not know how!

I well remember the consequences of operating too early. A thirteen-year-old girl came in towards midnight. The registrar thought she had early appendicitis. He wasn't certain, but felt that if delayed he would almost certainly be called out of his bed a few hours later, so he decided to proceed. A normal appendix was removed at a routine operation. However, for reasons that remain a mystery to me, but possibly due to a major break in aseptic technique, which seems unlikely, or a leak at the stump site, this poor girl developed, in sequence, a wound abscess, a pelvic abscess requiring drainage via the rectum, a subphrenic abscess and pleural effusion, once again needing surgery. She was in hospital for three months and I predict was almost certainly subsequently sterile. All because a man who became a medical missionary wanted an undisturbed night.

In the days before intensive care units, the sickest or most intensively monitored patients were either next to Sister's desk or in a cubicle. On one occasion, beside Sister was a retired army major who was confused following an aorto-femoral graft. This was something that sometimes occurred, especially with a borderline alcoholic. It was late at night and I was at the other extreme end of the ward when someone shouted, 'Doc, quick!' I looked up to see the major standing upright on his bed, naked. Despite the long incision in his abdomen, with three deft moves he pulled out in turn his stomach tube, intravenous drip and finally his urinary catheter with the retaining balloon still inflated. I received a mild bollocking the following day because the Largactil I administered to sedate him dropped his blood pressure and the boss was afraid that the graft might clot off.

Many patients requested sleeping tablets as a routine. Although it was easy in Nightingale wards to observe patients, there was little privacy and they could be disturbed by their fellows as well as necessary clinical activity, especially on reception nights. The standard preparations were barbiturates, the commonest being Soneryl by tradename. One morning, a chap asked me what I had prescribed for him the previous night and asked could he have it again, please.

'Certainly,' I said. 'Did you get a good night's sleep?'

'Not a wink,' he said. 'But the ceiling was covered in the most amazing colours and shapes swirling around like a kaleidoscope.'

The first ruptured aneurysm I saw was an old man who had collapsed on a trolleybus close to the hospital. He presented with diarrhoea but was also semi-conscious. Sometimes when an aneurysm leaks, the extravasated blood tracks down the left side of the abdomen behind the large bowel and strips it up, irritating it. However, this sort of presentation is rare. Anyhow, he was very lucky because the correct diagnosis was made rapidly. Just as the anaesthetist placed the mask over his face, the patient gasped, 'I wish to go privately', which later everyone found amusing. In fact, he belonged to a rare, if not unique, breed in Newcastle, being a stockbroker. He underwent a stormy passage and then recovery. When he saw Peter D in his consulting room six weeks after discharge, he apparently tossed his chequebook onto the desk and commanded his saviour—on this earth, at least—to fill out any sum he desired.

Visiting hours for relatives and friends were restricted and flexibility was available only in special circumstances, so that relatives could only discuss a patient's condition with seniors by appointment. But a notice in the corridor did state that a doctor would be available on a Saturday afternoon and that doctor was the houseman on duty. We dreaded it because the queue outside the room was like that outside a fish and chip shop on Good Friday. We were treated with respect and courtesy by almost all visitors, despite our youth, unlike the aggression one witnesses these days, by what mistakenly believes itself to be a more sophisticated and educated society.

Chapter Fourteen

In the Blood

As Christmas approached, rehearsals took place for the annual concert performed by the housemen and nursing colleagues. It never ceases to amaze me that successive houses unfailingly produced such an event. There always seemed to be enough people able to write, play musical instruments, dance, sing and operate lighting, all the requirements essential for a good production. John Davison wrote most of the sketches and his girlfriend and future wife, June Varley, choreographed the nurses. One sketch, *High Boon*, a parody of the film *High Noon*, featured Dr Tom Boon, the physician, and I played Crazy Legs Fleming, an Indian brave. My character was modelled on the previously referenced first assistant in surgery, Brian Fleming, who was a flamboyant character with campish mannerisms, including, as a chain smoker, tapping out his ash with an exaggerated gesture. I was able to play on this and made it my last move as I 'died' on stage during a gunfight. Although it was an affectionate portrayal, his attitude was ambivalent. He was flattered to be considered important enough to be included, but always seemed to hold something against his 'double'.

It all seemed a bit of a shambles, perhaps because it was difficult to get everyone together for rehearsals, what with doctors being on call and nurses having changing shifts. So naturally we were delighted when the Sunday afternoon dress rehearsal, with an audience of ambulant and wheelchair-bound patients, went like a dream. It was usual for the patients to be so receptive because it relieved their boredom and they liked to recognise staff they knew.

The audience for the three nights was drawn from hospital staff, including nurses and medical students. There was a consultant anaesthetist on the staff called Philip Ayre, who retired the year we qualified but continued to work intermittently and offered his services as a make-up artist for the concert. Dr Ayre was known nationally because of his invention in 1937 of a device called a T-piece for paediatric anaesthesia. It's still used in modified form and although he invented it, he admitted to not knowing how it worked. Born in 1901, he had the misfortune to have a harelip and cleft palate, which was understandably badly repaired at that time. As well as a severe facial deformity, he had a severe speech impediment and was difficult to understand. To top it all he was totally bald and presumably because as a young man he had been ginger, he masked his baldness with the most obvious carrot-orange wig. However, he was a kind man with a good sense of humour and was well-loved. At one of these concerts the audience responded poorly to the performance and Dr Ayre could contain his frustration from the wings no longer and bounded onto the stage doing a jig, tore off his wig, and shouted, 'Laugh, you buggers, laugh!', although it came out as, 'Raff, you muggers, raff!'

On Christmas Day itself, all hands were on deck. We were just at the transition from meals being served out individually from the various portions by sister-in-charge to where they were sent from the kitchen ready prepared, kept hot and covered, and doled out. For years, it had been a tradition for the senior consultant to carve the turkey and for medical staff to bring their wives and children into the hospital. Peter D wasn't going to let this custom die and Mrs D ferried my colleague Bill Cookson and I to the Dickinson abode only a few minutes away. Then, from two large ovens, we extracted two enormous birds so that he and Mac could do the honours.

All patients who were deemed safe to discharge were let out for Christmas, but the wards were still quite full because some patients could jump the queue if they were prepared to have their elective surgery—for example, a hernia repair or ulcer op—and still be an in-patient over the holiday. Many were agreeable, especially the self-employed. Hernias were in for five days but are mostly treated as day cases now.

Another long-standing festive tradition was the nurses touring the wards singing carols on Christmas Eve. They wore their cloaks turned inside-out to reveal the scarlet interior and were led by bearers of lanterns on poles. My wife, Mary, says nurses on duty did not always appreciate this custom because so many patients needed consoling after witnessing such a moving performance.

***

At the end of January 1967, we swapped surgery for medicine. John McCollum joined Bill Cookson and me for two months each on wards thirteen (male) and ten (female) on the first floor of the main block and two months in the Accident Room, assessing potential admissions and casual patients.

The consultants in charge were Dr RB Thompson, who had an interest in haematology and wrote a standard textbook on the subject, Dr AR Horler, a general physician, and six beds in an annexe were allotted to Dr DNS Kerr, a renal physician, who worked mainly on the Professorial Unit along the corridor.

The sister in charge of W13 was 'Bugger' Bell, the long-established nickname for an old-school spinster who was strict with both staff and patients, but fair. Grey-haired and plump with a determined brisk walk and attitude to match, hers was reputed to be the only ward Matron dare not visit if she was on duty. She was a stickler for bedmaking techniques and if a nurse made the slightest error, she would rip the sheets off in a temper and leave them dishevelled for the culprit to try again.

She was old enough to remember when nurses did the ward cleaning and polishing, after which they changed from the clothes appropriate for these tasks into their nursing uniforms at about mid-morning, when nurses now take a short coffee break. Keeping to custom, she ordered her nurses to 'Go to dress', which really meant, 'You can go for coffee now.'

Despite her severity she was well-liked by patients. In what was still a mining area, chronic chest disease was common, and some patients suffered recurrent admissions for acute flare-ups. In fact, each patient had a sputum jar as a routine and the amount of phlegm coughed up by some patients had to be seen to be believed. It was all kept for measuring and testing, and the jars were then washed and reused.

Sister Bell converted the far end of the ward into a games room, with a dart-board, small snooker table and cards table. On one morning blood round, I entered this annexe to find six of my victims, all dressed (Sister would not allow ambulant patients to wander around in pyjamas), and all cheerful, calling, 'Eyup, here comes Dracula' as I produced my syringes and bottles. I recall thinking as I made my way back down the ward that not one of those apparently fit and cheerful men would live much longer than three months. All had some form of acute leukaemia, myeloma or lymphoma, and such treatment as there was, certainly for adults, was largely ineffective. There was one drug, vincristine, which was useful in some patients. Over the years I have often described the following to my juniors and students as a lesson to be learned. One lady outpatient came for an injection of this drug and brought it with her, having called at the pharmacy on the way. It came as a powder in a glass vial and before injection had to be dissolved in sodium chloride (salt) solution. The 'blood trolley' had compartments for various glass vials, kept in meticulous order, and at one end were the ten-millilitre vials of 'normal saline'. I picked one out, scored the tip and broke it off, and inserted my needle ready to aspirate. I rotated the vial to see the label—and read, 'Potassium chloride 1gm'. Apart from the words themselves, the imprint was the same, with nothing to otherwise distinguish the two chemicals, and this vial had been placed in the saline section. If I had used that vial and injected it, the lady would have died of a heart attack on the end of the needle. Potassium chloride is added to intravenous infusions of saline and dextrose and is perfectly safe when given over a period of hours but is lethal as a one-off injection. Never take anything for granted. Always, *always* read the label.

The renal patients were nearly all young adults. All I did was carry out instructions. David Kerr had a devoted staff who performed all the difficult tasks. Haemodialysis was confined to a small selected number, largely on account of cost. The Newcastle transplant programme was in its infancy. Some patients were treated with peritoneal dialysis, in which the toxic products caused by renal failure are washed out using lavage with a fluid introduced into the abdomen via a permanent catheter, but complications in the form of infection and blockage were common. Those patients on the verge of needing dialysis were treated with a low protein diet and fluid restriction—the Giovanetti diet—which was burdensome.

I remember having to tell relatives that their loved one was 'not suitable' for dialysis. I almost believed it myself, it had been so drummed into me, but in truth they were perfectly suitable, there was simply an absence of funds. At that time, the treatment of such patients was undoubtedly worse than in some countries that we British looked down on in other respects. However, I was young and politically naive, and no doubt the situation was probably more complex than it seemed to me. Capital punishment for murder had only been abolished in 1965 and I distinctly remember a senior politician pontificating that no civilised nation should be seen to be executing criminals. And yet in my eyes convicted murderers were being kept alive, fed and watered at great expense, while we condemned otherwise innocent young people to death because the money was in the wrong budget.

One evening the renal SHO came onto the ward, literally in tears. He had just left a young woman whom he knew well and who was being treated for renal failure by peritoneal dialysis. She had suffered a series of complications, and this one was the final straw. Despite all his pleadings she refused admission, knowing she was returning home to certain death within a few days, but her mind was made up.

Before seeing any new patient, David Kerr sent out an extensive, structured questionnaire to be completed and produced at their appointment. A pale, thin little woman was admitted from Whitehaven in Cumbria for consideration of home dialysis. I don't think there were any satellite units for haemodialysis then, so I assume the alternative was dialysis three times a week in Newcastle, each session of about eight hours, plus a six-hour round trip! Home dialysis was expensive and tricky, requiring a prefabricated building on-site and a certain degree of intelligence. At the end of Kerr's missive were questions specific to members of each sex. In answer to, 'Do you have a discharge from your penis?' this poor woman answered, 'Yes'. I have sometimes wondered if that single reply had a negative impact on her destiny.

Alan Horler was a very good all-round physician and a serious man who did everything by the book and had little small talk at work. He referred to all students as 'Doctor', whether in jest or to reassure patients I never knew. Once, when demonstrating a full examination of the nervous system, he sent a student to obtain test tubes of cold and warm water, to assess sensation. The water from the warm tap was so hot that the student returned holding the tube at its rim. Horler grasped both tubes and winced slightly.

'Doctor,' he said. 'We are testing temperature, not trauma!'

Another time, he was teaching students and encouraging a stroke victim to unclasp the grip in his hand by gently prising open his fingers when he discovered a turd. The man was without speech and partially paralysed. The poor chap had somehow managed to grasp one of his own turds and in his confused state, had held onto it. Horler was totally unfazed and simply excused himself while he adjourned to wash.

During one ward round, a lady who was a patient of another consultant called him by name over to her bed and asked if he remembered her. Very politely, he replied that he did, that he would never forget her, and how was she after all these years? Doing well, she said, and although she was blind in one eye after the operation, she was now happily married with two children. After a short conversation, he returned to us with a mixed expression of puzzlement and amusement on his face.

'Easter, 1947,' he recalled. 'I was Dr Nathaniel Armstrong's senior registrar. She was a brittle insulin-dependent diabetic whom we could not control. Finally, in desperation, we asked Mr Rowbotham, a neurosurgeon, to do a hypophysec-tomy.'

The purpose of this procedure was to abolish the secretion of growth hormone, which is diabetogenic, from the pituitary gland. If successful, it would have also left her infertile, as she would no longer ovulate. Presumably, instead of the pi-tuitary stalk, the surgeon divided one side of the optic chiasma, which surrounds the gland, hence the combination of blindness in one eye and retention of fertility. Despite this, her diabetes seemed to have not done too badly, after all, twenty years on!

One of our haematology patients was an elderly man with aplastic anaemia, a condition in which the bone marrow fails, and the patient can be kept alive only with repeated blood transfusions. Previous housemen had warned us about this man, in that he was a nightmare to get a drip into. He was a pleasant but apprehensive chap who understood that his life depended on what remaining veins he possessed. Sister Bell knew the score and as soon as he arrived she would wrap him up in bed with several hot-water bottles to vasodilate him and make any remaining veins more prominent. Once the drip was in, he could maintain an almost catatonic trance to minimise movement that might dislodge it. I once gave him four pints via a small vein on the back of one thumb! He had received so many units of blood in the past that his skin was pigmented browner than the deepest suntan, caused by the breakdown of iron pigment in red blood cells. Fortunately, during our stint as housemen there was a transition from metal needles and glass syringes that were re-sterilisable to disposable plastic cannulas, tubing, and blood bags.

A middle-aged man was admitted with cirrhosis of the liver and ascites (a large volume of fluid in the abdomen). He was a cheerful, polite man whom I took to immediately. When asked if there was anything significant in his past history, he replied,

'Just a touch of syphilis during the war, Doctor.'

It was the 'touch' that amused me.

One afternoon I was chatting to our frequently put-upon psychiatrist, Dr Duffy, a Glaswegian with the typical sense of humour of that city's inhabitants, when our cirrhotic's 'partner' breezed past with a cheerful, 'Good afternoon, Doctor'. (Actually, we never used the word 'partner' then, we were more likely to say 'common law wife').

'Did you see that?' said Duffy, animated.

'What?' I replied.

'The triad—the saddle nose, the Hutchinson's teeth and the Glasgow accent—and the most important of the three is the Glasgow accent!'

I had met the woman a few times and noticed what I took to be a previously broken nose, probably as a result of a domestic dispute. The Glasgow accent was Duffy's amusing addition, but I hadn't observed the notched incisor teeth, which together with the collapsed bridge of the 'saddle' nose were characteristic of congenital syphilis. She had previously expressed concern to me that if her partner's condition was terminal, would we please keep him in until the end because she had woken up to find two husbands lying dead next to her and didn't want there to be a third. Despite her remarks, I was touched by the degree of tenderness and affection they displayed towards one another. He did not seem terminal, although cirrhosis is irreversible and progresses to liver failure, treatable only by transplant, which was unknown then. However, Easter Saturday started with flurries of snow showers but changed towards noon to bright sunshine lighting up the ward, made cheerful by vases of daffodils. I was directly opposite him when he glanced at the sky.

'It'll be a lovely day for Gosforth races, Doctor,' he said, and promptly died, just like cowboys and soldiers in Hollywood films.

Dr Horler admitted a retired GP in his eighties who was asthmatic and in the habit of injecting himself through his trousers with 1mg of adrenaline when he got an acute attack. His practice was in north Northumberland. I went to introduce myself and commented on his unusual surname of Trevor-Roper. I asked if he was any relation of Hugh Trevor-Roper (later Lord Dacre), the Professor of Modern History at Oxford.

'He's my son,' he said.

So I asked if he was related to Patrick Trevor-Roper, the eminent ophthalmologist.

'He's my other son,' came the reply.

So that was me put in my place.

The registrar was Mike Shaw, who had just returned to hospital medicine from general practice. He was a perfectly pleasant, handsome chap, bright-eyed and enthusiastic and somewhat hyperactive. But he was always demanding to know the results of investigations not long after they had been requested, and was keen to delegate but not so keen to do any work himself. He once instructed me to conduct an exchange transfusion on a woman who was entering hepatic coma due to fulminating hepatitis B. The condition was then called serum hepatitis to distinguish it from infectious hepatitis, now known as A. At that time, the only discernible difference was the incubation period, much longer for B than A, and the prognosis. Whereas infectious hepatitis (probably better known to the public as 'yellow jaundice', although all jaundice is yellow) was unpleasant, it was rarely fatal. Serum hepatitis had a significant mortality. As the name implies, it was spread by blood-to-blood contact, or by the sexual route. The source of infection in our patient was believed to be the inadequate sterilisation of a reusable needle in a community clinic where she received gold injections for her rheumatoid arthritis. It was known that the virus could survive exposure to various antiseptics in low concentrations and these were used where steam autoclaving was unavailable. The responsible virus was identified in 1965 and the following year the presence of the disease could be confirmed by an antigen in the blood, then called Australia antigen. The overall mortality is said to be 1-4% but for some unknown reason, some local epidemics have carried a much higher rate. I had seen a man die on Pav 1 who had received a transfusion elsewhere and was admitted to us with what was initially thought to be jaundice of 'surgical' origin, for example, gallstones or cancer of the pancreas, but in whom the newly available Australia antigen was positive. This present patient was clearly dying and the exchange transfusion was a forlorn hope. On the other hand, as the person carrying it out, I was at significant risk myself. It was slightly unnerving to see how far from the proceedings Mike Shaw kept himself. The whole performance took an inordinate length of time, as I was withdrawing blood from one arm and transfusing donor blood in the other. I think eight to twelve units were used, with no discernible benefit.

Tragically, between June 1969 and August 1970, the Edinburgh transplant unit experienced an epidemic. Twenty-six dialysis patients caught the disease, of whom seven died. In addition, twelve members of staff were affected, of whom four died, two surgeons and two technicians. Five years later, Colin Douglas wrote a bestselling novel, *The Houseman's Tale*, and although hilarious and sexy, it has a dark side to it, being based on the tragic events at Edinburgh where Douglas was a student and house physician at the time. Closer to home, a Sunderland surgeon pricked himself during an operation, caught the disease and recovered, only to develop aggressive cirrhosis within two years, which killed him.

Another tragedy was a woman in her thirties who was twenty-something weeks pregnant with her second child and developed what appeared to be a superficial thrombophlebitis of one thigh. Although painful, this in itself was no cause for concern, but if the thrombus spread to the deep veins in the groin where the superficial vein joins them it can cause a life-threatening pulmonary embolus, as previously described. She was fit and able to walk in, but as a precaution was given an anticoagulating dose of intravenous heparin, which required monitoring as an inpatient. Had she not been pregnant she would have been started on warfarin tablets and could have gone home, but warfarin crosses the placental barrier, with the risk of fatal bleeding in the foetus. After a few days of heparin, she became restive and anxious to return home to her eleven-year-old daughter and husband, so the drip was stopped, and she was told she could go. Elated, she got up to pack—and fell down with a pulmonary embolus. Despite all efforts, she died a few hours later. She had gone into labour, but the foetus was too small to survive and had not been born at her death.

I mentioned earlier that we only had half the beds on W10, the other half being shared by Tom Boon and Reg Hall, the whizz-kid endocrinologist who later became Professor of Medicine at Cardiff. Reg admitted patients with acromegaly for a full three weeks of investigations. Acromegaly is a condition caused by oversecretion of growth hormone from a benign tumour of the pituitary gland, which sits in a little hollow in the base of the brain. If the tumour appears before the bones fuse, it leads to gigantism. If it appears afterwards, the hands and feet become broadened and spade-like and the facial features enlarged. Some organs such as the heart also enlarge, but this is not beneficial. When the condition is well advanced, it is a spot diagnosis (that is, recognisable on sight). Anyway, the only reason for raising the subject is that, having spent three weeks undertaking blood tests, glucose-tolerance tests and x-rays, as these ladies were wheeled out of the ward upon discharge, they unfailingly called out their thanks to all and sundry. This tended to cause wry amusement to the cynics among us, because nothing had been done that was directly to their benefit, and most would not receive any treatment, because none was truly effective.

***

We saw plenty of overdoses. Most were females and did not seem to be determined attempts at suicide, more likely gestural of 'a cry for help'. The only fatal overdose I witnessed was that of a female lab technician who had an argument with a senior colleague and immediately swallowed potassium cyanide from a jar on the shelf. This occurred in the hospital's basement, but she was brought up to the Accident Room where attempts to reverse the actions of the lethal substance were predictably unsuccessful.

Coal gas was still prevalent and I remember a couple of cases, unconscious, with the typical cherry-red faces (carbon monoxide adheres preferentially to haemoglobin in blood cells, preventing oxygen from so doing, but despite the sufferer being hypoxic, the carboxyhaemoglobin in the blood maintains a 'healthy' red appearance). They were treated with inhalation of a mixture of oxygen and carbon dioxide, ready-made in cylinders specific for this purpose. The CO_2 stimulates respiration so that victims would over breathe, in the hope that eventually the oxygen would displace the carbon monoxide.

Most overdoses were of various tablets—aspirin, paracetamol, antidepressants, barbiturates etc. If they were conscious, and I cannot remember any who weren't, they had a stomach washout as a routine, using a wide-bore rubber tube. If they were uncooperative, they were held down by as many people as could be mustered. Technically, this was assault and would not be tolerated today, but I cannot recall any retribution. In retrospect, the procedure may have been useless, as any absorption would probably have already occurred. I sometimes think it was more of a deterrent to try to prevent a recurrence.

One sad case was brought in on a sunny Saturday afternoon by a couple of Good Samaritans who were crossing the Town Moor when they encountered a nineteen-year-old youth quite calmly sitting on the grass feeding himself tablet after tablet from the largest bottle of aspirin either of them had ever seen. When admitted, he was perfectly cooperative. The story was that he had been diagnosed as schizophrenic. Both parents were on the academic staff of Durham University. One day, presumably as the result of a paranoid delusion, he assaulted his mother and was thrown out of the house and left to fend for himself. He was a clean, tidy and obviously intelligent lad, but the present overdose had been provoked by an incident in Leazes Park. While sitting on a seat at one end of the lake, he observed a boxer dog at the far end, standing on the surface of the water and then dashing towards him, all the while growing larger and larger until it finally bounded up to him, by this time gigantic, and started licking his face. He told me that when the vision faded and he calmed down, he realised this was his first visual hallucination, that such experiences were indicative of a poor prognosis for his condition, and hence he had better end it all before he deteriorated. He therefore wandered off to Boots for a large bottle of aspirin—no questions asked then, not like the third-degree you experience these days when you want some Lemsip for a cold. I have often wondered over the years about this young man and what became of him.

Most of the attempted suicides were admitted overnight and consultation forms sent to the duty psychiatrist. Mary and I were due to be married on April 1st (I know, I know, but it was after midday), and about a week before, a couple of young women were admitted, each following an overdose. I provided some background information, together with my version of their underlying motives, to our old friend, the Glaswegian Dr Duffy. After spending some time with them, Duffy approached me in an expressionless manner.

'They tell me you're getting married next week. Is that true?' he said.

'Yes,' I confirmed with a smile, thinking he was going to congratulate me.

'That's a pity,' said he, looking sad. 'Because you know absolutely nothing about women. The one you feel sorry for is a psychopath who should be discharged immediately. She might do it again, but at least you won't be on call tonight if she does. The other, the one you think is putting it on, deserves all the support we can give her.'

Chapter Fifteen

Wedding Bells

Despite my shortcomings when it came to the opposite sex, the wedding went ahead anyway, at 2pm in St Peter's Church, Bishopton, an idyllic location with a village green, two pubs and the mound of a Norman castle nearby.

Although Roger was my best friend throughout medical school, his house jobs had been at Shotley Bridge and Sunderland, and we hardly saw one another, so John Davison agreed to be best man, with Roger and Gerry as groomsmen. Her father having died when she was only seventeen, Maurice, her sister Ann's husband, gave Mary away.

Everything went well, but as I expect with most people, it was all a bit of a blur and only snippets remain in the memory. We honeymooned in Majorca at a small resort, C'an Pastilla, which I gather has since burgeoned in size. It was the first time I had travelled abroad or flown. We entrained at Darlington, where a clergyman ushered us into his carriage, having seen our party of well-wishers throwing confetti. From King's Cross we went to the West London Air Terminal on Cromwell Road to go by bus to Heathrow. A noisy old Trident took us to Barcelona, where we waited some time for the short flight to Palma. This was in what a typically loud American lady described as, 'Gee, we're not going in that little old thing with a windmill on the front, are we?' that is, a propeller-driven aircraft. We passed through customs manned by serious-looking, mustachioed Civil Guards with coal scuttles on their heads, who chalked our cases wearing their most intimidating, yet somehow comical, expressions (remember, this was still the time of Franco). We were then met by a man with a *wagons-lit* emblem on his epaulettes, who beckoned me.

'Senor Clarke?'

'How does he know who I am?' I asked Mary.

'Because you're the only daft so-and-so wearing an overcoat,' was the morale-boosting reply.

We spent two happy weeks at what was a quiet resort, certainly in the spring. The only unfortunate occurrences were, firstly, being of a sensitive nature, I persuaded my bride that as we were in Spain, we must watch a bullfight. This was a serious mistake. Secondly, being a practical man, the film in my Olympus Pen-EE camera failed to wind on and none of our honeymoon photos came out.

We returned to a rented flat in Holly Avenue, Jesmond, although for another three months I was on call alternate nights. The next anxiety was the interview for the Anatomy Demonstrator posts, the prelude to a surgical career. In fact, I cannot remember any interview and wonder, in retrospect, if they judged us on our applications, although years later, Prof Scothorne told me we were all 'hand-picked'. Those few who applied but failed to be appointed were recommended to his old boss, Prof Wyburn of Glasgow, and obtained jobs there. One of them, Graham Teasdale, who I mentioned as a fellow guest at the Great Gatsby party in Guisborough, went on to become Professor of Neurosurgery in Glasgow, and a knight of the realm. Another, my fellow houseman, Bill Cookson, gave up the idea of surgery in this country and emigrated to Alberta, where he combined general practice with such basic surgery as he felt capable of, in the town of Drayton Valley's small hospital, ninety minutes by road from the nearest city, Edmonton.

There was the usual party on the last night of the house, but I remember it as something of an anti-climax, and the following morning we went our separate ways, me to start married life properly. My mother paid for me to have some driving lessons in Hull during the few breaks I had as a student, but I had not yet passed my test. However, with what savings we had, we bought a Mini from a fellow houseman. While some of my fellow demonstrators could drive and hence could do GP locums during the summer, I started the job early and was employed at preparing dissections for the new students coming in October. Meanwhile, Mary accompanied me with my L-plates and I took some more lessons, but I failed the driving test the first time, so it was several months before I could drive independently.

I was one of nine demonstrators. The sole female, the lovely Jeanne Bell, was sidelined to oversee the anatomical and physiological teaching for the speech therapists, themselves all female. The remaining eight of us taught anatomy to the new medical students. Each of us supervised twelve students, divided between two dissecting tables.

Two of my group had local connections, but the rest were from the south of England. One girl was the daughter of Dizzy Dawes, an ENT surgeon in the RVI, and one boy, Rob Clay, was the great-grandson of an eminent surgeon from the 1920s onwards who had died aged ninety-two, when Rob was eleven. He remembered the old boy having an artificial leg. Both Miss Dawes and Rob were highly intelligent, but both soon gave up medicine, the former for mathematics, and Rob for I don't know what. It makes one wonder whether they initially did what was expected of them but were disappointed. Another girl was Jennifer Crawley, who naturally became known as 'Creepy' Crawley. She was very short-sighted and in one anatomy session she positioned the male cadaver with his legs flexed and separated so that she could sit at the end of the table, with her glasses perched on the end of her nose, dissecting his perineum. She obviously realised what a sight she was to behold, for she suddenly said in a slow, southern drawl, 'If ownly my mather could see me naawh!'

It was six years since my colleagues and I had been in the position of our new charges, and the 'new' curriculum was now well established. Sensibly, the students had far less dissection to do themselves, and the syllabus was completed within the year because we demonstrators prepared 'parts' at different stages for them to study in conjunction with their textbooks. I was interested to find that as long ago as 1893, the eminent Newcastle surgeon, George Grey-Turner, said that too much time was spent in the dissecting room, to the point that it became recreational. Out of six students, only two could be put to use at any one time, creating a tendency for them to chat and wander around. Socialisation of students can be beneficial in education, but one can overdo it.

Although we had previously covered the whole of human anatomy, it was a long time ago, and the purpose of the present job was the symbiotic one of both teaching the subject and revising at the same time. Often, we were only a few steps ahead of our students. The Primary FRCS exam we would take the following summer was in both anatomy and physiology, and we had to revise the latter from the books. Our old physiology books by Bell, Davidson and Scarborough (BDS) and Samson (Sammy) Wright, standards for decades, were no longer recommended by the powers that be and were replaced by Ganong or Guyton, both Americans. Generally speaking, I have not been a fan of American medical books, mainly because they tend to be overwritten and concentrate more on the abstruse at the expense of more mundane but important topics. However, Ganong served my purpose. *Gray's Anatomy*, the long-established textbook the title of which is familiar even to the lay public, is really a reference book that deals with systems rather than regions, such as arteries, veins and nerves, rather than the arm, pelvis etc. Over the years I have acquired several anatomy books, which begs the question, 'Why?' when anatomy is a finite subject, or, if you like 'dead', in that there is no more left to learn. Well, some are purely anatomical. Others include what is termed 'applied' or 'surgical', demonstrating anatomy pertinent to certain pathological conditions and operations. Some include anatomical variations, and all have different forms of illustration so that several two-dimensional views can help the student gain a better three-dimensional impression of a difficult area in the absence of a dissected specimen or cadaver. I dare say that with modern computerised images, the anatomy textbook is on the verge of extinction. I have one called *Extensile Exposure* by Arnold Henry, an Irish surgeon, which is written in the most beautiful English. Ian McNeill said it was in blank verse, I presume as an intellectual exercise for the author. The book generally recommended for the Primary was R J Last's *Anatomy: Regional and Applied*.

Being graduates of an English medical school, by tradition and expectation, we took the Fellowship of the Royal College of Surgeons of England (FRCS Eng). It was in two parts, the primary, in the preclinical sciences, and the final, in clinical, pathological and operative surgery. In the UK there were two other Fellowship diplomas, those of Edinburgh and Glasgow, but whereas the finals of all three colleges were completely separate, the respective primaries were interchangeable, and in Newcastle it was traditional to go to Glasgow. Naturally, graduates of the Scottish medical schools, of which there were five in my day, took either (or sometimes both) Edinburgh or Glasgow. The overseas graduates, for whom I always felt sorry, because as far as the primary was concerned they were at considerable disadvantage compared to those of us fortunate to be demonstrators, sometimes took a couple of Fellowship exams as an insurance, treating the process as a lottery, although the fees were not inconsiderable. In England, it was always said that the Edinburgh exam was easier than the English. This may have been chauvinism, but I cannot recall hearing of anyone failing Edinburgh and passing London, whereas the opposite situation was quite common. Perhaps the Scottish examiners were kinder. Few, apart from local graduates of that fine city, took the Glasgow final by choice—I don't know why. Of course, in Scotland there has long been an Edinburgh-Glasgow rivalry, permeating almost every field of endeavour. Frankly, apart from individual prejudices, in theory it mattered little which FRCS one passed, as long as one did so eventually. As far as I am aware, there was no limit to the number of attempts a candidate could make, the exams being held every six months, but a trainee would not obtain promotion to senior registrar, or in some cases, registrar, without the full diploma. In the privileged position of teaching anatomy full-time, with free evenings and weekends, the pressure was on us to pass, as we were expected to do. And fortunately, so we did.

Virtually the final act of the year was to play in the traditional staff versus students cricket match. Our captain was the senior lecturer, Tom Barlow, the genial, cricket-mad Mancunian, who opened the batting for his village of Ponteland until he was sixty and was well-loved by generations of students. He called us for a practice session, during which, I regret to say, I performed quite well. I was a good all-rounder at school but had not indulged in any sport since. I say regret because based on this session I was well up in the batting order and also asked to bowl. Not only was I out first ball, but in my single over one of my own students, a cocky lad who had opened the batting for his public school, hit me all over the ground. We were relying on the fast bowling of John Reid, a physiology lecturer who had skittled our team out in the same fixture when we were the students. Sadly, although still accurate, John's pace had much reduced in the intervening five years. This seemed unbelievable, but only four years later he appeared on the ward at Shotley Bridge where I was by then a registrar, for a coronary bypass. He was still only in his early thirties and, tragically, only survived another year or two.

***

It was during this year that Aunty Elsie died. She developed cancer of the pancreas, for which she had either an exploratory operation or a bypass (I'm unsure whether she was jaundiced or not). I last saw her with Mary in her home at 29th Avenue, North Hull Estate, shortly after her surgery, when she gave me a weak, sad smile and asked if there was anything that could be done 'about the pancreas'. The C-word was never mentioned. 'Uncle' Stan, Elsie's only child, had moved to Waltham, near Grimsby, to lecture at the local polytechnic on electronics rather than continue as an electrician for the NEB, and Elsie moved there to be nursed during the last three months of her life. As usual, my mother did her duty, and visited every weekend, taking the Humber paddle-steamer ferry and then the bus—and she was still working at the time. Elsie was the eldest and shortest-lived of the three sisters, dying in her seventieth year. I remember her with great affection, for she was kindness personified, although like so many others she had a mundane, sad life.

***

The next move up the greasy pole of life was to obtain a surgical trainee post. In theory, one perused the Appointments section of the *British Medical Journal* and applied accordingly. In practice, the attitude was to either make a break early, ie now, or try to stay within the catchment area of one's own teaching hospital. Some people, usually single, are wanderlusts and are happy to move from job to job, and those who succeed in reaching the top are either very talented or very lucky. At some point, one needs to stay in a place long enough for whatever skills and ability one possesses to become recognised, and hence a culture of parochialism developed, especially with competition being so fierce. This may seem unhealthy, but most contracts were for six-to-twelve months and the prospect of moving around the country at such frequent intervals, especially when married, was not an attractive one. Fortunately for me, and others, two of the nine demonstrators decided to stay in the anatomy department, one did obstetrics, one initially entered ENT and Jeanne Bell carried on with the speech therapists, thus reducing the competition in general surgery.

A registrar post came up on the surgical rotation. This was a marvellous opportunity, but daunting, because although it provided a three-year contract comprising one year's general surgery and four six-month specialties, it was a promotion above our capabilities, as we had practised no clinical medicine since our house jobs. However, one had to apply for it, and Mike Black and I did. Fortunately, Mike, who was far more confident than me, got the job, but then I was recalled and asked if I would be happy with a temporary appointment as senior house officer (SHO) to Mr GY Feggetter. This position was one step below registrar in seniority. Well, I jumped at the chance. I hadn't been a student of his and had only come across him during final year revision clinics, but he was revered as one of the hardest and fastest working surgeons in the hospital. Typical quotes—'Time's money', and 'Come on, I do it the same way every time.' He was also approaching retirement and I might not get the chance to work for him again. The word 'temporary' did not deter me, although it did appear on my payslip, which was somewhat unnerving.

Chapter Sixteen

A Year with Feg

Thus began one of the most enjoyable and formative years of my surgical life. George Young Feggetter was about sixty-three when I began working for him, but he looked older. He was of average height, with silver hair, a wispy moustache and a slight stoop. He had piercing blue eyes, which occasionally twinkled with amusement, because he only rarely smiled and one hardly ever saw his teeth when he spoke, which he did in a sort of flat monotone, with little emotion, and always to the point. He had little small talk and disliked wasting time. Despite this somewhat unexciting description, he appeared to be in total control and exuded confidence. He looked the part and was a surgeon's surgeon, a man you would trust with your life and, more importantly, the lives of your close family.

One small example was when Ian Massey, an ex-Feg houseman and now a registrar at the Westminster Hospital with Prof Harold Ellis, developed a hernia and came all the way to Newcastle to have it repaired by Feg. A couple of us called on him in his cubicle. He was beaming and commented on how Feg's presence simply filled him with confidence and strength—and all he had was a hernia! But when it's your hernia, it's the most important thing of the moment. Feg was a man of his time, one whose whole life revolved around surgery, with little interest in research, and who, despite his well-disguised humanity, communicated only briefly with patients. He knew what he was doing, told patients what he planned to do, and left the details of informed consent to us. Patients just had to look at him to put their trust in him.

'Don't tell them anything, Clarke,' he once said to me. 'They always get it wrong.'

To modern ears, this sounds brutal and typical of the public perception of the arrogant, uncaring consultant surgeon. But he was none of these, and I knew exactly what he meant, which was that if you are going to give a detailed explanation, make sure they fully understand.

Such a failure in communication embarrassed me once, when I found that a man admitted as a perforated ulcer actually had Bornholm disease, a self-limiting viral infection that causes short-lived but painful inflammation of the lower intercostal muscles and diaphragm. It is named after an island off Denmark where an epidemic occurred. Jokingly, after reassuring the patient that he did not need an operation and would soon be well, I casually mentioned that the condition used to be referred to as 'The devil's grip'. The following morning, a group of anxious relatives collared me at the entrance to the ward demanding to know what this terrible devil's grip condition was. Served me right!

GYF's surgical and military career was distinguished. He was registrar to the internationally famous surgeon, George Grey-Turner, a Newcastle graduate who left the RVI to become the foundation Professor of Surgery at the Royal Post-graduate Medical School at the Hammersmith Hospital in London in 1935. During the war he served with the army in the 69th General Hospital in the North African desert, reaching the rank of brigadier and accumulating a wealth of experience in operating under difficult conditions, usually using spinal anaes-thesia administered by himself. It was a rare day when he omitted to make any reference to either when he was Grey-Turner's registrar or when he was in the desert. Grey-Turner himself was a disciple of James Rutherford Morison, another Newcastle legend, who started as a GP in Hartlepool and ended up Professor of Surgery in the RVI. An Edinburgh graduate, Morison worked for Lord Lister, the man who pioneered antisepsis in surgery using carbolic acid air sprays, swabs and wound dressings, and who made surgery much safer than before, despite initial opposition. So those of us who worked for Feg proudly traced our surgical genealogy via him, G G-T and Morison, to Lister himself, hoping that some trace of these great men would rub off on us.

When I started in August 1968 the senior registrar was Nick Batey, who had been a demonstrator to us in 1962-3, and David Muckle was the rotating registrar. Nick had left Newcastle for Glasgow and Greenock, returning as SR at a much younger age than usual. He had a big influence on me because we later also worked together on the Dickinson/McNeill unit. In those days, junior trainees learned more operative surgery from the SR than the consultants. Subsequently, Nick became a consultant at Noble's Hospital in the Isle of Man. Dave Muckle had an interest in the fledgling subject of sports medicine and wrote a book on it. While in Newcastle he offered his services firstly to Whitley Bay AFC and then, as his career in orthopaedic surgery progressed, to Oxford United when the club was owned by Robert Maxwell and Dave was at the Radcliffe Infirmary. Finally, at Middlesbrough FC, he became a medical advisor to the game's governing body FIFA, possibly aided by the influence of Harold Shepherdson, one-time Boro trainer and loyal assistant manager to Sir Alf Ramsay for England's 1966 World Cup victory.

For new members of staff, Feg set a military example by including an inspection of the toilets in his ward round. On one occasion after a patient fell, he checked the slippers of everyone on the wards!

The male ward, W4, had a charge nurse, Johnnie Ritson, assisted by Ella Hall, a junior sister, and before Feg's morning rounds they would flit from patient to patient asking if they had eaten breakfast and what. At some point, sure enough, Feg would enquire if someone had eaten, but he aimed the question at the nurse, not the patient. It was simply a way of keeping them on their toes.

He was also keen on bowel movements, and on one occasion Johnnie Ritson answered such an enquiry by cupping his hands and slowly rotating his fingers as if moulding potter's clay and replied that the patient had passed a well-formed motion with 'a rosy glow', with such a look of satisfaction on his face that it could have been one of his own. Not a flicker from Feg. Then he turned and said, *sotto voce*, 'Ritson's job is beds and bowels, and between you and me, he's not so hot on the bowels.' Slightly unfair, and unusually disloyal of him, and perhaps a rare, clumsy attempt at humour. He usually treated the staff with courtesy and respect and supported them if in difficulties. When he did a large bowel resection he routinely expected preoperative preparation with washouts, to reduce the incidence of infection and anastomotic breakdown. My wife was a student nurse on the unit and was summoned to theatre, having been given the message that Mr Feggetter wished to see the nurse who prepared the patient's bowel. Naturally, she hurried along in trepidation, only to be complimented on one of the cleanest bowels he had seen in a long time.

Although he was respected, he wasn't feared. I never saw him lose his temper, although he could be irritated and would use sarcasm, but not in a nasty way. He once asked a houseman to perform a task that he either forgot to do or had insufficient time. He then requested the result and was informed that it had not yet been completed.

'Oh, I'm sorry, I wouldn't ask you to do something you felt incapable of doing,' he muttered, not unpleasantly.

With the houseman quietly seething, Mr Feggeter of course knew it would now be done promptly, and it was.

Outpatients ran on Saturdays from 8.30am until 12.30 or 1pm. He taught students on every case and clearly believed that a grilling was educational and/or good for the soul, for although students arriving late were welcome to stay and learn by observation, he never questioned them, reserving that privilege for the early birds. On a Sunday morning there was a leisurely ward round, followed by the ritual of the SR accompanying him to his car, while I suspect he probed him about any problems.

At operating lists on a Monday and Wednesday, knife was expected to be applied to skin at 8.45am precisely. When I was with him there were two anaesthetists, Johnnie Wheldon and Joan Millar. One Monday morning, just as Feg's scalpel touched the abdomen, the patient's right hand came up under the green covers. Feg backed off and after a short while, the same thing happened again. Feg turned to Johnnie Wheldon.

'If that happens again, I'm going to give him the scalpel and he can do the operation himself,' he said.

Joan Millar was a lovely woman and a genuine lady, kind and elegant with beautiful manners, cheerful, friendly and a tower of strength—one of the old school, the like of which I fear we shall not see again. She was chiefly a cardiothoracic anaesthetist, and I later had the pleasure of working with her both at Shotley Bridge and when I was SR to Peter D and Mac. On one of her lists, Feg decided to do an open prostatectomy on an old man who was a Jehovah's Witness. Blood loss in such an operation can be unpredictable, but Feg promised the man he would not be transfused. However, Joan said that if he bled badly, she would do so, which created a bit of an argy-bargy.

'No, no, you mustn't,' said Feg. 'I've promised him'—with the nearest hint of panic in his voice that I ever witnessed.

'Well,' said Joan, 'he's a stupid old man and you're a stupid old man for promising him.'

Of course, internally we were all creasing ourselves because we had never heard him addressed in this way before.

'Your attitude to him,' replied Feg, 'is like that of the Romans to the early Christians.'

Anyway, all went well, and they did not transfuse him, but his blood count did drop during the next few days.

Feg could more or less tackle anything, but his main interests were peptic ulcer surgery and urology. About ten per cent of the population suffered from peptic ulceration at some stage in their lives. I will digress somewhat on the history of treatment for the condition. Symptoms tend to be episodic, and in the main, controlled by diet and antacids, but a minority had either frequent attacks of severe pain or complications such as bleeding or perforation, which could be life-threatening. It was known that reduction or abolition of acid secretion by the stomach could heal duodenal ulcers, the commonest type, affecting the first part of the duodenum just beyond the outlet of the stomach, but oddly enough, the quantity of acid secreted by sufferers was usually the same as the rest of the population. Because the precise cause of the ulceration remained unknown until the 1980s, and no drug treatment was curative, surgical attempts to heal ulcers concentrated upon abolition of acid secretion.

Historically, the first and obvious method was to remove part of the stomach (gastrectomy). Most acid is secreted by the upper two-thirds, the lower third secreting a hormone, gastrin, which also promotes acid secretion in response to food. To achieve adequate reduction in secretion, both the lower third and a significant portion of the upper two-thirds had to be removed. Because the resulting gap was too great to join the remnant of the stomach to the duodenum, the latter was closed off and a loop of small bowel just beyond the 'C' of the duodenum hitched up and stitched to what was left of the stomach. This operation had a small but significant mortality in the early days, and some patients, although relieved of their pain, had debilitating complications such as anaemia, weight loss and what is sometimes known as 'dumping' after meals, when food passes rapidly through the stomach causing nausea, sweating, dizziness and diarrhoea. Of course, once performed, this procedure could not be reversed, so you can see that whereas those who suffered life-threatening complications were candidates for surgery, those with symptoms alone had to 'earn' their operation—in other words, their suffering had to warrant the risks.

By chance it was found that bypassing a duodenal ulcer by joining a loop of small bowel just beyond the duodenum (so named because it is twelve finger-breadths long) to the stomach, could also heal an ulcer, but unfortunately about a third of patients later developed a so-called stomal ulcer at the join. This gastroenterostomy was a far less risky procedure, but the incidence of stomal ulcer soon made it obsolete. However, in the 1940s, American surgeon Lester Dragstedt pioneered an operation to divide the vagus nerves just below the diaphragm. One function of the vagi, the tenth cranial nerves, is to stimulate acid secretion by the stomach, but they also control its motility, so that division of them has to be accompanied by pyloroplasty, a drainage procedure, in the form of slightly widening the pylorus, the small muscle at the outlet of the stomach.

When I worked for Feg, vagotomy and some form of drainage was the standard operation for duodenal ulcer. He always did a gastroenterostomy, rather than a pyloroplasty, his logic being that the former healed the ulcer, and the vagotomy prevented a stomal ulcer, a reverse take on the usual thought process. The problem with dividing the trunks of the vagus nerves, one in front of and one behind the lower oesophagus (gullet), was that it could be either incomplete, because of tiny additional strands of nerves that could be missed or could provoke disabling diarrhoea in a tiny number of patients. Hence, the next advance was highly selective vagotomy, which did not require drainage, as the branch of the nerve to the pylorus was preserved. Just as this operation was taking off, Sir James Black developed an H2-receptor antagonist drug to suppress acid, and this soon became 'the magic bullet'. Histamine released at the end of the vagus nerve directly provokes secretion of acid by the gastric cells but is not blocked by traditional antihistamines used for allergies such as hay fever, because the 'receptor' on the cell wall is of a different type (hence termed H2 to distinguish it from H1). It was this crucial difference that Black pounced upon and created the drug, initially cimetidine (trade name Tagamet), later superseded by similar 'cleaner' ones such as ranitidine (Zantac). This change occurred shortly after I became a consultant in 1980, and it was surprising how many patients still opted for surgery rather than taking tablets, even though side effects were absent or minimal. We did not yet know that long-term continuous medication with such drugs was safe.

At roughly the same time, an old-fashioned treatment made a resurgence. In Victorian times, bismuth salts were used for dyspepsia and a new preparation was released on the market. It worked perfectly well, the drawbacks being that it was in liquid form, which stained the tongue black, and it also smelled like a baby's nappy, being ammoniacal. It so happened that for the first nine months of my new consultant job I commuted from Newcastle to Middlesbrough, and I soon developed symptoms characteristic of a duodenal ulcer. I didn't disclose this to my new colleague, who specialised in gastrointestinal surgery, in case he insisted on passing an illuminated tube of liquorice down my throat to confirm the diagnosis, while gleefully sharpening his knife. Instead, I simply begged free samples of Tagamet or Zantac from the pharmaceutical reps who visited from time to time. My symptoms were always relieved, but returned promptly on finishing the supply. Finally, the bismuth preparation became available in tablet form and a delightful redheaded rep was only too keen to give me a plentiful supply because she was finding it difficult to make headway against the more convenient H2-receptor antagonists. Interestingly, I've had no trouble since. Then one of my ex-Newcastle colleagues, David Tweedle, a senior lecturer in Manchester, published a paper showing identical healing rates of ulcers by both H2-receptor antagonists and the bismuth preparation, but a much lower recurrence rate following cessation of the latter, which was unexplained. This suggested that despite its inconvenience—now simply reduced to making the tongue black—it was the better drug. Soon, the mystery was solved, when the Australians Marshall and Warner discovered that the bacterium, Helicobacter pylori, was present in the stomachs of a large percentage of the population, and in a significant minority was causing an inflammation that damaged cells, allowing stomach acid to create duodenal and gastric ulcers. The modern treatment therefore aims to eliminate the bacterium and explains why bismuth—which, like some other heavy metals such as mercury, lead and silver, has an antibacterial effect—reduced the rate of relapse in Tweedle's trial. However, as Einstein said, 'Everything should be made as simple as possible, but not simpler.' Despite this dramatic breakthrough, which appears to have made obsolete all those years of striving to heal ulcers by surgically reducing acid secretion, we still don't know why the majority of the population harbouring H. pylori in their stomachs remain completely asymptomatic.

But back to Feg. He did a vagotomy and gastroenterostomy on almost every list and by the mid-1960s had followed up 250 cases for at least fifteen years, a remarkable achievement considering the earliest he can have started the procedure was on his return from the war. His other interest was prostatic surgery, and he was the only surgeon in the RVI to use the resectoscope, an instrument inserted along the urethra to chip away an enlarged prostate gland and relieve obstruction.

Whereas specialist urologists can remove the whole gland from within its capsule, I think Feg was content to create what some derisorily call an 'English Channel', and if necessary, repeat the procedure if the patient developed recurrent symptoms in the future.

He treated many patients with superficial cancers of the bladder, so-called papillomas (inaccurately actually, because this term implies that they were benign), using diathermy and some excisions. Such patients were followed up with check cystoscopies every six months or so. I was amused once when he was looking through a cystoscope, only to turn away looking puzzled.

'I don't know this bladder,' he announced—and went to the top end to see if he recognised the anaesthetised man's face.

He asked me to review all his patients with cancer of the prostate, which meant obtaining records going back to 1946. Almost all had been diagnosed by rectal examination alone, in which the gland feels hard and irregular, but microscopic confirmation was only obtained in those who needed a resection to improve their urinary stream. The treatment at the time was the synthetic hormone Stilboestrol, which resembled the female oestrogen. Because, like breast cancer, that of the prostate is hormone dependent, it can be suppressed by opposing its driver, testosterone (that of the breast being oestrogen). But this came at the price of feminising side effects such as breast enlargement, loss of libido and shrinkage of the genitalia. I remember one man saying his penis was now so small that when he needed to get it out, he borrowed his wife's crochet hook! It was only years later that trials showed that although Stilboestrol did shrink the gland and often reduced metastatic spread also, overall its effects were to shorten life because its fluid-retaining properties pushed some elderly men into congestive heart failure. Fortunately, enormous strides have been made since then in the screening, diagnosis and management of prostate cancer.

Chapter Seventeen

Lumps and Bumps

As the SHO, I was responsible for the weekly outpatient theatre list—generally known as 'lumps and bumps'. I removed such as sebaceous cysts, lipomas and ingrowing toenails under local anaesthesia. I had some interesting experiences in spite of the relatively mundane tasks. I was removing a cyst from a lady's head when she suddenly spoke up.

'I've taken a liking to you,' she said. 'I should get you to tighten up my tush. My husband says it's like dragging a sausage down the back alley.'

I had never heard the noun before, although the meaning was unmistakable, and the only occasion since was in a film in which Burt Reynolds was referring to the rear end of a horse!

I also did the bougie clinic for usually elderly men who had a stricture of the urethra, from whatever cause. It was routine for them to turn up without an appointment when their stream began to slow down. The charge nurse, Stan Knott, who I got to know well over the years, kept a file of cards containing each patient's details. Men also came at intervals to have their hydroceles tapped. These collections of fluid in the scrotum which accumulate for no obvious reason are usually treated surgically in younger patients, in whom they are less common, but elderly men often preferred periodic tapping with a hollow cannula under a local.

Bougies are sterilised, curved metal instruments of graduated sizes, and are passed after instilling a local anaesthetic gel into the urethra. With skill, the bougie can be almost dropped into the bladder under its own weight, guided by the operator using a manoeuvre referred to as the *tour de maître*, but it needs some practice and when performed clumsily can cause intense discomfort in the region of the prostatic urethra. And so it transpired when I did my first one, the man jumping a bit on the table and cursing, before I eventually succeeded. I hadn't worn a mask, as one was unnecessary for the procedure, but I kept my eye out for the same man when he attended for his next dilatation several weeks later, when I made sure I wore one so that he would not recognise me. By this time I was an expert, and when I finished he let out a huge sigh of almost orgasmic satisfaction.

'By, that was great,' he said. 'Nothing like the bugger who did it last time!'

A man whom I did not recognise appeared one morning.

'How long have you had your stricture?' I asked.

'France, 1915,' he replied.

A war hero, I thought, soft-heartedly.

'Fractured pelvis, was it?' I queried.

A little smile appeared, together with a faraway look.

'No,' he said. 'Mademoiselle from Armentieres.'

I had momentarily forgotten that the commonest cause of urethral stricture was gonorrhoea, whereas in this clinic the majority followed transurethral resection of the prostate.

One man who wanted his hydrocele tapped came prepared with a Winchester-sized bottle, into which he requested transfer of the fluid. He maintained that it did his leeks the world of good! Competitive leek-growing was and still is a passion in the now ex-mining communities. The straw-coloured fluid contains a small amount of nitrogen-containing protein, so he was probably correct.

***

Stan Knott's first job in nursing was in the Wallsend shipyards, manning the accident unit. One day, he was summoned to the hold of a ship where a man had fallen. He was more used to dealing with cuts and foreign bodies in the eye, but this sounded serious, so he responded urgently. He found a chap groaning at the foot of some stairs, holding his side and complaining about his ribs. Stan had just knelt to examine him when he heard the bell of the ambulance, which belonged to the yard. The rear doors opened and men appeared from nowhere to fill it with wood, paint and other useful commodities. Stan was told to never say a word, or else, and with him, 'patient' and helpers aboard, the vehicle set off, bell clanging and the dock barrier raised in anticipation. Once clear, it began unloading its contents at intervals along the streets of Wallsend. As Stan said, an ambulance was the only vehicle allowed through the barrier without being searched.

'Mind you,' he added. 'All the houses were painted battleship grey!'

Although Feg often mentioned his time as Grey-Turner's registrar, I cannot recall any anecdotes about the great man. Joan Millar was probably the last person locally who could remember him, other than Feg himself, and she would have only been in her late twenties when he left for London in 1935. She described him as a small man with a large head, who still retained the accent of his native north Tyneside.

Apparently, he was parsimonious and shod his own boots using beaten tin cans. When the Mayo brothers, founders of the world-famous clinic named after them in Rochester, Minnesota, visited him in Newcastle, he sent his house surgeon down to the Haymarket, a few hundred yards from the hospital, to buy twenty Wills' Woodbines, placing them in a silver cigarette box on the lunch table as a gesture of hospitality. There was a longstanding rumour that as a young man, he learned to sleep only on alternate nights, using the time to write up his case notes. This must be a myth but conveys the undoubted zeal he possessed for surgery, as well as his own ambition. Certainly, he knew exactly when he was performing his 1,000th appendicectomy, so his records must have been meticulous. There is also a well-known story which I have heard independently from several sources, of a wealthy lady from Sunderland who was sent to him as a patient. But she didn't take to his appearance and accent and decided to try her luck in Harley Street, first consulting a local GP for a referral.

'Madam,' he is said to have promised, 'I shall send you to the best surgeon in England'—and promptly referred her to Grey-Turner.

GG, as Joan Millar called him, died aged seventy-four in 1951. The university created an annual lecture in his memory and at the last one I attended, in 2012, the distinguished American professor who gave it confessed that on receiving the invitation he had never heard of him.

Sadly, he is probably best remembered for the least of his contributions—the description of a discolouration in the flanks of some patients with severe pancreatitis, similar to bruising, caused by digestion of tissues behind the organ by its enzymes. Known as Grey-Turner's sign, it was originally described as a diagnostic tool, before the days of serum amylase estimations and CT scans, but it occurs late in the disease process and is of no help in management.

Feg didn't appear to be a fast surgeon, but he had great economy of movement and the clock didn't lie. He was the only surgeon I observed who tied knots with both hands. This went against all surgical tyros' inclinations. Had we not been taught to tie our shoelaces with both hands? Surgeons used only one, throwing and tightening the knot with it, while holding the forceps and needle taut in the other. It took ages to perfect the manoeuvre and be able to do it swiftly and safely (the latter being achieved by switching direction of hand movement so that the lie of each throw created a secure knot that would not slip). But if Feg saw you perform this act, he would deflate you by telling you to learn to tie a knot with both hands.

Whenever he operated on a hernia, the operation note was always headed, 'Radical cure of inguinal hernia'. It was a somewhat exaggerated claim I thought because everyone knew there was a small but significant risk of recurrence in the hands of any surgeon, as evidenced by the number of operations differing in varying degrees which were designed to prevent such an event. However, here was Feg copying his master, Grey-Turner, who, I discovered many years later, used exactly the same heading himself.

Operation notes were, and still are, either written by hand just after the procedure, in the restroom while waiting for the next patient to be anaesthetised, or spoken into a Dictaphone. Feg's were dictated, partly because his writing was verging on the illegible, but mainly because a typewritten copy was included in the notes and also kept to be bound together with all the others in a volume at the end of each year. Also included were summaries of each patient's admission, created after their discharge by the senior registrar, and also sent to the GP. Because the latter took a little time, the houseman gave each patient a handwritten discharge summary. This custom of a consultant surgeon having personal records of all his patients collected and bound is a bygone one in the NHS, like hot food out-of-hours, and dining rooms.

One brief operation note: a man from Cumbria was sent to Feg with an enormous neglected low-grade sarcoma, which no-one over there would tackle. It said, 'I dissected this man off his tumour'.

He had also spent time in Berlin, as well as the Hammersmith. I presume it was on a secondment, but I recall his comments on the differences in the plight of the unemployed during the Depression of the 1930s. He described the Geordies as sitting disconsolately 'on their honkers' at street corners, smoking with the so-called 'depression hold', preserving the life of their cigarette against the wind by cupping the hand around it, apparently apathetic. He compared them with his observations of their German counterparts, now that Hitler was Chancellor, marching in disciplined fashion like soldiers, with polished spades replacing rifles, digging ditches or undertaking similar projects in return for their meals or a small stipend.

He did have a private practice, but we were hardly involved. One patient, a certain Lady Frances X, had a minor procedure on a list. As she was recovering from the anaesthetic, a nurse gently tapped her cheek.

'Lady Frances, Lady Frances... wake up!' she repeated in a soft *faux-posh* voice.

She delivered the 'wake-up' component in a sing-song uplift. A few yards away, sitting in the surgeons' rest room, Feg gave a chuckle and revealed that 'Lady Frances' was not a real lady at all but was merely christened as such by her socially ambitious mother, who had prefixed her sister's name similarly to ease their passage through life.

Just after midnight one evening, I admitted a student nurse with obvious appendicitis. She was accompanied by her medical student boyfriend. Both lived away from home but the girl's family were insured with BUPA, so she insisted on private treatment. With slight trepidation, because I had never needed to call Feg before, I telephoned him and he answered almost as if he was waiting for the call, with no hint of having just been woken up. Within minutes he was in the Accident Room, hair slicked down and parted, immaculately dressed and bright as a button. Actually, I think he enjoyed the whole process—the satisfaction of being needed and of doing something he had probably not done for a long time, an appendicectomy. Fortunately, the culprit was acutely inflamed. The night staff were all of a fuss—a consultant's presence was an infrequent novelty. Most consultants had their little foibles and Feg was no exception. After removing the appendix and placing the purse string around the basal remnant, he turned round.

'Phenol,' he said.

My old friend Sister Joan Newbould was aghast. She had long forgotten that he dipped a scalpel in eighty per cent phenol to kill any bacteria sitting there, before inverting it out of sight. But it soon appeared, and I started doing this myself afterwards, as I thought it was such a good idea.

Miriam Stoppard was a houseman to Feg and has written that she, too, had a great admiration for him. It was said that she was once assisting him in theatre when one of her false eyelashes fell into the wound, and the rumour was that he vowed never to have a female house officer again. Possibly apocryphal, but within type on both accounts. Miriam was always glamorous.

One amusing incident occurred when a relative of his, a farmer, needed a bowel operation. He was in a cubicle and, for whatever reason, was slightly confused. Feg and retinue called in during a ward round and anxious to assess the state of his handiwork, he asked the patient if he had moved his bowels.

'Well George,' he said—the George was prolonged and accentuated and this alone was amusing to us, never having heard him addressed by his forename before. 'Well, George, I thought I was going to, but all I could do was faaarrt'—this last word drawn out to emphasise his degree of frustration, which left everyone, Feg excluded, struggling to suppress laughter.

Feg simply turned to his houseman.

'What have you got him on?' he said, which begged the question but was correctly interpreted as what he had prescribed for sedation.

'Soneryl, sir, 200 milligrams, three times daily.'

'You should be doing your house job in the Royal Homeopathic Hospital,' he said.

'Double it, and if that doesn't work, double it again. I don't want to hear another peep out of him.'

Another private patient was the author Catherine Cookson, whom I had not heard of at the time. She had a rare condition called hereditary haemorrhagic telangiectasia, in which multiple abnormalities of tiny blood vessels occur throughout the body, and when present in the stomach and intestine can periodically bleed and cause anaemia. Before the advent of flexible endoscopy, all that one could do, and usually all that was necessary, was to observe until bleeding ceased and transfuse as required. During one admission, the houseman was Keith Baxby, originally from Sheffield. Some years later, Keith's wife was reading a Cookson novel when she exclaimed that one of the characters was a Dr Baxby. Keith grabbed the book and commented that she must have named the character after him because he was aware how uncommon his surname was, with only a handful in the Sheffield telephone directory.

Feg took both me and his son, Jeremy, to the annual dinner of the North of England Surgical Society. Jeremy was a classmate of mine and succeeded me as his father's SHO. It sounds nepotistic (except that I believe this term refers to an uncle's influence, rather than a father's), but I genuinely believe Feg thought he was the only one who could inculcate a solid surgical background into his son. He was not an arrogant man, but had enormous self-belief. This event occurred before the time when spouses were invited. After the meal, people were circulating, and a Middlesbrough surgeon, 'Sam' Mottershead (he was actually Sidney, but his initials were SAM), approached and thumped Feg on the back.

'Now then, George, I've got plans for thee,' he said.

The 'thee' derived from his still-evident broad Lancashire accent, Mottershead being a contemporary of the anatomist, Tom Barlow, and both devotees of Prof Wood-Jones, the eminent professor of anatomy at Manchester. We receded into the background, while Sam bemoaned in a monologue how he had been treated unfairly and deserved to be in a teaching hospital rather than sent down the road to Middlesbrough like a missionary to treat the natives. The target of his invective was Professor Bentley, the first professor of surgery appointed by the university, who held the post between 1945 and 1954, when he emigrated to the USA, possibly to mitigate alimony costs (even in those days).

'Bentley never liked me,' said Sam, who proceeded to destroy the man's character, piece by piece, all the while Feg and his young guests remaining silent, but me at least, being both astonished and entertained.

Finally, came the *coup de grâce*.

'And do you know, George, do you know, that bugger cribbed to get through his matric?' he declared. 'And I know that for a fact, because my wife's sister sat next to 'im in class at Bolton.'

And with that, he sat back with his arms folded and a look of triumphant satisfaction on his face.

After a few seconds' pause, Feg replied.

'Well, I'm sorry you feel that way, Mottershead, because I know that Professor Bentley had a soft spot for you.'

'Did 'ee really, George? Did 'ee really? Well, I always said 'is bark were worse than 'is bite. And you know, 'ee were a talented pianist. There were more to 'im than met the eye.'

And gradually, bit by bit, he began to retract it all. This was one of those occasions when Feg's eyes twinkled and there was just the hint of a smile. Ever since, I have thought this was a lesson in life for two young men from a wise old one.

By strange coincidence, I replaced Sam at Middlesbrough when he retired, like him, begrudgingly, feeling that I was not one of the Professor's 'boys'. But somewhat belatedly I was reassured when Prof's secretary told me he once said he should never have allowed me to leave Newcastle for Middlesbrough.

At some point during the year, Feg requested a new-fangled operating table made in Sweden, which transferred the patient from trolley to table without human assistance. Unfortunately, the table was not mobile and had to be embedded into the floor via its cylindrical support. When they built the RVI, the theatre floors were of terrazzo, installed by craftsmen from Italy. No expense spared in 1906. So the central area of the theatre had to be dug out and the tiny tiles painstakingly replaced around the support. After several months of continual malfunction, Feg admitted defeat, and it had to be removed.

'I might have known,'' was his bitter comment .'After all, the Swedes sold ball bearings to the Germans during the war.'

Feg's theatre and the professorial theatre were on the same section of corridor, and each had two sets of doors. When both were open, one could walk past and see directly into the theatre. While this table was being removed, a workman was excavating the base with a pneumatic drill and both doors were open, making him visible from the corridor. Phillip Ayre, the anaesthetist with the cleft palate and speech impediment, was making his way past having just completed a list in the prof's theatre. He gazed in and saw dust flying from the base and heard the racket from the drill.

'What's up?' he asked. 'Has George Feggetter lost a swab?'

***

During this year Grandma Barmby died. She outlived her eldest daughter, and I never knew how she reacted to this loss. At times in my presence, she occasionally showed signs of confusion, especially at our house and when it was late, but she seemed capable of an independent life, with my mother's support. One Sunday afternoon, the man who lived opposite her in Ivy Terrace was shaving before going out, and in his mirror he spotted flames in Grandma's front window. In her confusion she had set the curtains alight. Thank the Lord that neighbours

retrieved Grandma. My mother was sent for, the fire brigade extinguished the
fire and the police contacted the family doctor, who admitted poor Grandma to
the De la Pole Hospital, formerly the asylum, alongside other victims of senile
dementia, for the blessedly short duration of three months, before her death at
the age of eighty-eight. I managed to get the day off for the cremation but was
advised by my mother and Aunty Flo not to view her body beforehand as 'she
doesn't look very nice'. So now there were just the two sisters from my mother's
side of the family left in Hull.

***

About halfway through the year, Nick Batey and Dave Muckle moved on, to be
replaced by George Abouna and Dave Lambert. One could write a book about
George—in fact, someone has, although it is somewhat one-sided. Dave Lambert
was a tall, blond, blue-eyed, handsome ladies' man (meant in the nicest possible
way). Apart from an attraction to the opposite sex, he was an experienced Alpinist
and was preparing to join Chris Bonington's expedition to climb Annapurna as
the team doctor.

Unfortunately for Dave, Abouna got up to his old tricks as a control freak,
trying to manage every aspect of Dave's surgical life, when he should have been
allowed his head and own judgement. Eventually, his frustration boiled over into
conflict, which Feg was given the task of trying to resolve. He interviewed everyone involved in private, but it was mere guesswork among the non-combatants
as to what action, if any, was agreed. Before all this blew up, Dave and I got on
well. He was amused, but delighted, at Chris Bonington's public relations skills
in achieving sponsorship for the expedition and said that in nearly every post he
received a different gift, including a watch, sweater and boots. He had spent a year
in Ibadan, Nigeria, doing pathology, which did not seem the usual pathway for
a budding surgeon, but Dave had a wanderlust. Just before I left the unit, Dave
fell one weekend while climbing near Keighley. He was transferred from Airedale
hospital a few days later with a fractured pelvis and fractures of the wrist and

head of the radius of one arm. Feg either knew about Dave's reputation with the ladies or perhaps it was his urological interest, but he seemed obsessed about the integrity of his urethra because of the pelvic fracture. He was also concussed and confused, especially at night, so much so that cot-sides were erected to prevent him getting out of bed, although they didn't stop him trying to cuddle any nurse who came close. I next saw him in my new job in orthopaedics when PN Robson sorted out his arm by excising the head of radius and internal fixation of the wrist. In typical fashion, PNR stuck out his prognathic chin before the operation.

'You would have thought that if he wants to be a surgeon,' he said, 'he would have had the good sense to land on his feet.'

I cannot recall the time periods involved, but instead of wrecking his trip to Annapurna, he recovered sufficiently well to be sent out with the initial party, which included Don Whillans, an experienced climber but nonetheless a nasty piece of work, who even as a companion, insulted Dave as being a hanger-on. In fact, Dave did climb as far as expected, but on his return, as well as having an amazing suntan, he had lost about two stones in weight. Unfortunately, he was forced to curtail a career in operative surgery because of premature arthritis of the wrist and the last I heard, he was a consultant in rehabilitation in Seattle and was married, naturally, to a nurse.

I had a wonderful year working for George Feggetter, learning a lot and gaining confidence. He was an excellent role model and I hope some of this is revealed in the memories I have selected. I have concentrated more on anecdotes than clinical work, but the latter was fulfilling, and I look back on the experience with gratitude and affection.

One case I will describe in detail was tragic, interesting and rare. I operated on a woman in her fifties who had a perforated appendix. It was almost *de rigeur* with a perforation in those days to drain the area, so that if pus developed subsequently, it would find its way out along the drain, rather than causing an abscess in the abdomen. The briefly fashionable antibiotic of the day was intravenous tetrex, a long-acting tetracycline, and because a paralytic ileus might be expected in such a case, the stomach was decompressed by a nasogastric tube, and she was given intravenous fluids. Two days postoperatively, I found her in a cubicle because she was confused in the night and moved from the general ward. Now, postoperative confusion does not occur in this age group unless there is a potentially serious cause, such as sepsis or a respiratory problem, and I was worried. Soon she developed severe watery diarrhoea.

'She hasn't got staphylococcal enterocolitis, has she?' asked Feg.

Sure enough, a stool specimen showed staphylococci, and her antibiotic was therefore switched to cloxacillin, which should have been effective, but to our amazement, the organism was resistant. She developed renal failure, went rapidly downhill, and died. Naturally, I was very upset and racked my brains as to where the fatal organism could have originated. A staff nurse reminded me that on her first night she had been next to a 'boarder', that is, a patient transferred temporarily from another ward to vacate a bed for an emergency. This had been an old woman with a discharging, infected wound following a plating of a fractured femur. I discovered that her infecting organism was a staphylococcus sensitive only to chloramphenicol, which somehow had been transferred to my patient, probably via the nasogastric tube. When I informed the bacteriologists they could not have shown less interest if they tried, a reaction which shocked me even then, for long before MRSA, the so-called 'hospital staph' was already rearing its head.

***

As my year as an SHO approached its end, a couple of jobs became available on the three-year registrar rotation, designed to provide the incumbents with the requirements for the final FRCS diploma examination. Both George Bone and I were appointed. All such rotations incorporated a year of general surgery on one of the four units in the RVI, six months accident and emergency, six months orthopaedics, and two six-month sub-specialties from four possibilities—urology, cardiothoracics, plastics and neurosurgery. It was a matter of chance which specific rotation one was allocated to, depending on the precise vacancy. As it happened, George and I both started on orthopaedics. Not only was there tremendous relief at obtaining such a post in my own teaching hospital, but the security of a three-year contract meant Mary and I could leave our rented flat in Jesmond and try to find a place of our own. We were quite happy in the flat, except that early on we had a burglary, and the thieves stole many of our wedding presents. Mary worked as a staff midwife at Newcastle General, where she was happy, doing plenty of deliveries and being so pleasant to the many Asian mothers from the west end that she was nicknamed (and referred to as) Nurse Khan by one of the senior midwives who was not so enamoured, complaining that they spoke no English except 'Maternity Benefit'.

Before we married Mary had been 'on the district', which meant getting used to the relative poverty of Byker and Scotswood Road. She remembers one woman who was about to give birth when she wanted to move her bowels. The WC was down the back stairs and they didn't have a chamber pot, so the expectant mother requested a large saucepan from the kitchen.

'Don't worry, Norse, aal give it a good scour out afterwards,' she promised.

Some expectant fathers fancied their chances with the midwife, seeing it as an opportunity for the sexual congress they had been missing. They seemed to believe that a soiled white vest and a combined aroma of sweat and beer was irresistible. One woman actually told Mary to beware of her husband. Another said that after her last pregnancy her husband removed her perineal stitches because they were pricking him as he attempted what was very premature intercourse.

At work, the midwives were also pestered by the sort of sexual harassment that would not be tolerated today. One Greek registrar was apparently forever pushing himself up against potential victims, sometimes obviously aroused, to use the euphemism, until the aforementioned senior midwife rebuffed him.

'You're like a dog in heat,' she said.

Being compared to a dog was perceived as such an insult that he never tried anything again.

The Mini started to let us down, so we bought a second-hand Fiat 850, and when that did the same we purchased our first new car, a Fiat 600. At the end of our street in Jesmond was the Cradlewell pub, run by ex-Newcastle United and

Scotland winger Bobby Mitchell, a pleasant man, and his equally delightful wife, with her blonde beehive hairdo. We went there nearly every Sunday evening if I was not on call, but one had to arrive by 7.30pm to guarantee a seat in the lounge. Bobby and his wife were almost always on duty together on a weekend, but on one occasion he was absent. Usually, when the door opens in a pub, the occupants of the room turn to see the newcomer. At about 8pm the door opened and in strolled Bobby, in suit and tie, followed closely behind by the Tyneside legend and former Newcastle United and England footballer Jackie Milburn. There was silence for a few seconds before a succession of voices from all parts of the room, all making the same offer in the same way.

'It's Wor Jackie, it's Wor Jackie! Here, Jackie, thor's a seat here, thor's a seat here.'

Our hero just smiled and looked embarrassed. Like Bobby, he was a modest man with no airs and graces, despite his fame.

With the security of my three-year contract, Mary and I began looking for a place of our own, and found a two-bedroomed bungalow in Wideopen, just off the Great North Road, almost merging with the mining village of Seaton Burn to the north, and an easy drive to the RVI and Newcastle city centre.

And so began the next stage of my surgical journey. There was no guarantee of success. Several targets were necessary and one could still fall at the next or final hurdle. I had to obtain the final part of the FRCS diploma, become competent in a wide variety of operative procedures and probably complete a research project before reaching the hallowed post of senior registrar, the prelude to a consultant post. All this seemed to be a long way ahead at this stage. However, I did eventually achieve my goal, but now I was returning to orthopaedics, my previous experience of which had not been an entirely happy one. Hence, it was with the usual combination of excitement and apprehension that everyone experiences, not confined to medicine, that comes with promotion and a new challenge.

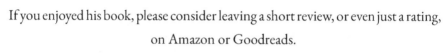

If you enjoyed his book, please consider leaving a short review, or even just a rating, on Amazon or Goodreads.

Reviews make a huge difference for independent publishers and by writing one you will be helping David's books reach a bigger audience. Thanks so much!

Coming soon

Opening Up Volume 2: My Life in Theatre

Email mcgearymedia@gmail.com marking OPENING UP in the subject line to be added to our mailing list and we'll inform you as soon as Volume 2 is available.

Also Published by McGeary Media

The Man Behind the Mask

The autobiography of a pioneering orthopaedic surgeon

John Anderson transformed thousands of lives by replacing bringing hip and knee replacement to the North-East of England.

The son of a steelworker, he was inspired to follow a career in medicine after spending a year in hospital with a back condition as a teenager and seeing the surgeons transforming patients' lives.

In the 12 years before he started work in Middlesbrough in 1975 there had been 12 knee and hip replacements carried out by surgeons trying out the emerging techniques.

By the end of his career John was doing that many each week. A founder member of the National Joint Registry Committee, in 2004 he was awarded the CBE for services to medicine, receiving the award from the Queen at Buckingham Palace. John was loved and respected by everyone who had the privilege of knowing him, whether as a friend, patient or colleague.

This is the story of *The Man Behind The Mask*.

Available on Amazon in ebook and paperback editions or read for free on Kindle Unlimited

Married to the Man who Washed Himself Away
A Memoir of Motherhood, Marriage and Obsessive Behaviour

Born into a working-class community in the North-East of England, Joan longs for a stable home and children of her own and marries handsome older man Kenny.

But marital bliss is short-lived—Kenny's obsessive behaviour makes a normal family life impossible.

When tragedy strikes, the couple's fragile bond is shattered forever, and Joan realises she must escape her troubled marriage if she is to find the happiness she craves.

This is Joan's story – a moving account of heartbreak, resilience and love.

Available on Amazon in ebook and print editions or read for free on Kindle Unlimited.